Mathematics in Mind

The monographs and occasional textbooks published in this series tap directly into the kinds of themes, research findings, and general professional activities of the Fields Cognitive Science Network, which brings together mathematicians, philosophers, and cognitive scientists to explore the question of the nature of mathematics and how it is learned from various interdisciplinary angles.

The series will cover the following complementary themes and conceptualizations:

Connections between mathematical modeling and artificial intelligence research; math cognition and symbolism, annotation, and other semiotic processes; and mathematical discovery and cultural processes, including technological systems that guide the thrust of cognitive and social evolution

Mathematics, cognition, and computer science, focusing on the nature of logic and rules in artificial and mental systems

The historical context of any topic that involves how mathematical thinking emerged, focusing on archeological and philological evidence

Other thematic areas that have implications for the study of math and mind, including ideas from disciplines such as philosophy and linguistics

The question of the nature of mathematics is actually an empirical question that can best be investigated with various disciplinary tools, involving diverse types of hypotheses, testing procedures, and derived theoretical interpretations. This series aims to address questions of mathematics as a unique type of human conceptual system versus sharing neural systems with other faculties, whether it is a species-specific trait or exists in some other form in other species, what structures (if any) are shared by mathematics, language, and more.

Data and new results related to such questions are being collected and published in various peer-reviewed academic journals. Amongst other things, data and results have profound implications for the teaching and learning of mathematics. The objective is based on the premise that mathematics, like language, is inherently interpretive and explorative at once. In this sense, the inherent goal is a hermeneutical one, attempting to explore and understand a phenomenon—mathematics—from as many scientific and humanistic angles as possible.

More information about this series at http://www.springer.com/series/15543

Marcel Danesi

Ahmes' Legacy

Puzzles and the Mathematical Mind

 Springer

Marcel Danesi
Department of Anthropology
University of Toronto
Toronto, ON, Canada

ISSN 2522-5405   ISSN 2522-5413   (electronic)
Mathematics in Mind
ISBN 978-3-030-06621-5   ISBN 978-3-319-93254-5   (eBook)
https://doi.org/10.1007/978-3-319-93254-5

This Springer imprint is published by the registered company Springer International Publishing AG part of Springer Nature.
The registered company address is: Gewerbestrasse 11, 6330 Cham, Switzerland

Preface

A good puzzle, like virtue, is its own reward.—Henry E. Dudeney (1857–1930)

Puzzles emerged at the dawn of history, and perhaps even before that (Olivastro 1993), putting on display the powers of the human imagination to grasp truths in its own playful way. There is no culture without riddles, no era of time without clever conundrums that are intended to defy logic, and no faculty of mind, from language to visual perception, that has not produced its own ludic artifacts. In mathematics, puzzles have often played critical roles in the history of the discipline, above all else as miniature investigative models of inherent principles or patterns of hidden structure, often leading to significant discoveries. For instance, Alcuin's River Crossing Puzzle or Euler's Königsberg Bridges Puzzle contained, respectively, the blueprints for combinatorics and graph theory. Even one of the oldest mathematics texts, the Egyptian *Ahmes Papyrus*, which dates back to before 1650 BCE, turns out to be essentially a collection of puzzles that were designed to illustrate specific mathematical ideas and to emphasize the power and simple beauty of mathematics—the same beauty of which the Pythagoreans spoke and which they saw reflected in music and in the movements of cosmic bodies alike.

This book will attempt to argue that the origins of some, if not many, mathematical concepts originate in the form of suggestive puzzles. Although this is well known on the part of mathematicians, and even though the branch of "recreational mathematics" provides a specific locus in which to examine individual puzzles and their theoretical implications, there are few overall general psychological discussions of the significance of puzzles as revelatory manifestations of how the mathematical mind works, as far as I can tell. I have not designed this book as a rigorous treatment of puzzles within the domain of recreational mathematicians, but rather as an excursion into the relation of puzzles to mathematical discovery, and what this relation tells us about the brain. No particular technical knowledge is thus required, since I will discuss each and every puzzle in a general way. My objective throughout is to show how some of the classic puzzles of mathematical history are essentially imaginative explorations of quantity and space in ludic form. Needless to say, so many ingenious puzzles have been invented that it would be brazenly presumptuous to claim that I have chosen the most important or paradigmatic ones here. Moreover, most have already been studied in depth as sources shaping significant parts of mathematical history. My aim is to take a plausible look, so to say, inside the mind of the puzzler, both as maker and solver, in order to trace the source of puzzles within the mind. This book is not a historiography of puzzles, which I attempted to treat schematically elsewhere (Danesi 2002), although the histories of some puzzle genres will be discussed whenever they are relevant. Rather, I have written it to convey the intriguing psychological story that puzzles tell us about the mathematical imagination.

As mentioned, in no way do I intend to imply that mathematicians are unaware of the cognitive value of studying puzzles. Indeed, they discuss them, write about them, and research them in meaningful ways. So, this book is really an overview, a summing-up, or synopsis, of the role of puzzles in the history of mathematics. I have discussed the ideas in this book with students in a course I have been teaching for over a decade at Victoria College at the University of Toronto called "Puzzles, Discovery, and the Human Imagination." I am very grateful to them for all their insights over the years. I am also grateful to Elizabeth Loew at Springer for giving me the opportunity to put my ideas down in writing, and to Dahlia Fisch for her expert editorial advice and encouragement. Any infelicities that this book contains are my sole responsibility. I sincerely hope that mathematicians and general readers alike will glean something from it that might have escaped their attention, and thus provide an "interesting" perspective on mathematical discovery. I truly believe that the systematic study of puzzles is as important as the study of any human construct or mode of creative expression. As the great modern-day mathematician, David Hilbert, once observed, "Mathematics is a game played according to certain rules with meaningless marks on paper" (cited in Pillis and Rose 1988). The creative source of those marks is often to be found in math games themselves.

Toronto, ON   Marcel Danesi
2018

Contents

List of Figures

List of Tables

Chapter 1
Puzzles and Mathematics

Pure mathematics is, in its way, the poetry of logical ideas

—Albert Einstein (1879–1955)

The English word *puzzle* covers a broad range of meanings, alluding to everything from riddles and crosswords to Sudoku, optical illusions, and difficult conundrums in advanced mathematics. As a generic categorical label, it is a convenient one for classifying diverse manifestations of what is arguably a singular psychological phenomenon, which can be called the *ludic* mind, that is, a mind that grasps or models ideas through some form of creative intellectual play. The word *puzzle* was first used to describe a game in a forgotten book published by Robert Dudley and Abram Kendall around 1595, titled *The Voyage of Robert Dudley Afterwards Styled Earl of Warwick & Leicester and Duke of Northumberland* (Warner 2015). It derives from the Middle English word *poselen* "to bewilder, confuse," a definition that certainly can be applied to most of the classic mathematical puzzles.

It was not until the invention of the jigsaw puzzle around 1760 by the cartographer John Spilsbury that the term started migrating to other areas of ludic creativity. But its entrenchment into language and groupthink was a slow and gradual process. Even Lewis Carroll (1880) in the nineteenth century referred to his classic "puzzles" as "problems." It was toward the latter part of the nineteenth century that the term *puzzle* became a common one as professional puzzlists, such as Sam Loyd (1914) and Henry E. Dudeney (1917), started using it to designate their clever creations, many of which had implications for mathematics.

A specific word for this concept did not exist in any of the languages of the ancient world (as far as can be told). So, labeling some ancient "problem" as a "puzzle" constitutes a retrospective form of reference. Riddles, on the other hand, were identified and named concretely as distinct playful verbal artifacts by the languages and cultures of antiquity. Today, we tend to classify riddles as types of puzzles. But in antiquity, they were perceived as something vastly different. They were seen as manifestations of divinatory speech, and thus as possessing prophetic value. The concept of "problem" did exist in the ancient languages, surfacing initially in the domain of geometry, where it was used to refer to a shape or figure that had to be

© Springer International Publishing AG, part of Springer Nature 2018

M. Danesi, *Ahmes' Legacy*, Mathematics in Mind,

https://doi.org/10.1007/978-3-319-93254-5_1

constructed in some preestablished way. From this, the word was extended to cover any mathematical (geometrical or arithmetical) task that required a specific kind of method in search of a solution. Among the ancient "problems," it is easy to see with modern eyes that some were actually "puzzles." It was Plato who first extended usage of the notion of *problema* beyond the domain of geometry in several of his works (for example, Plato 2004), applying it to describe any difficult situation that needed some type of resolution. Even a cursory glance at ancient mathematical texts will show, retrospectively, that emerging ideas in mathematics and philosophy were explored via "problem" formats that were essentially "puzzles." Psychologically, the distinction is a crucial one. A problem can be characterized as a presentation of some situation based on a question and answer (Q&A) format providing all the information needed to reach the answer directly; a puzzle, on the other hand, gives information that appears to be incomplete or intractable, thus making it much more difficult to reach the required answer. We know, however, that there is an answer at the end and that the "fun," so to speak, is getting to it. Scott Kim (2016), in fact, defines a puzzle simply, yet insightfully, as follows: "A puzzle is fun, and it has a right answer."

The purpose of this chapter is to discuss, in an initial way, why puzzles are important in the history of mathematics, revealing how the mathematical mind manifests itself in their actual forms and contents. This is an extraordinary claim, of course, so it will be pursued and examined throughout this book in some detail. Anecdotal-historical support for it comes from the fact that the intellectual explorations of ancient mathematicians took the form of puzzles. These are to the history of mathematics what symphonies are to the history of music. They are mirrors into the minds of mathematicians and composers respectively.

Problems, Puzzles, and Games

It is useful to differentiate conceptually among four terms, *riddles, problems, puzzles*, and *games*, given especially that these are often used interchangeably, both within and outside of mathematics. Riddles are ludic verbal artifacts, which ask a question framed in such a way that the answer is not easily discernible. One of the oldest known riddles (if not the oldest) is the so-called Riddle of the Sphinx:

What creature walks on all fours at dawn, two at midday, and three at twilight?

The legend behind this riddle reveals how the ancients viewed riddles—as prophetic forms of speech. The mythical Sphinx posed the riddle to anyone daring to enter the city of Thebes, which it guarded night and day. Those visitors who were incapable of solving the riddle paid for their ineptitude with their lives at the hands of the monster. As the story goes, it was Oedipus, son of Jocasta and Laius, queen and king of the very city that the Sphinx enslaved, who solved the riddle by answering *"human being*," the only creature on earth that does indeed crawl on all fours in infancy (the dawn of life), walks on two legs as a grown-up (the midday of life), and on three, with the help of a cane, in old age (the twilight of life). Because he answered correctly, Oedipus unwittingly fulfilled the sinister prophecy that he tried so desperately to avoid, as the well-known legend recounts. Oedipus had been left to die as a child on a mountain by Laius, because he had been told by an oracle that he would be killed by his own son. The infant Oedipus was, however, saved by a shepherd who raised him and eventually told him about the prophecy. On his way to Thebes to unravel the mystery of his birth, Oedipus killed a man in a duel. For getting rid of the monstrous Sphinx by solving the riddle, the Thebans made Oedipus their king to replace Laius who, by tragic irony, was the man Oedipus had killed on his journey to Thebes. As the new king he married the widowed Jocasta, who was, of course, his mother. On discovering what he had done, Oedipus put out his own eyes in a fit of madness, and Jocasta hanged herself.

Not all stories about riddles have such an ominous shade to them. Riddling was also a common type of recreational game at feasts, celebrations, and intellectual competitions. The biblical kings Solomon and Hiram, for example, organized riddle contests simply for the pleasure of outwitting each other. The Greeks used riddles not only for prophetic reasons, but also as part of their leisure activities, especially at banquets, as we might do even today at social gatherings. The Romans made riddles a central feature of the Saturnalia, a feast that they celebrated over the winter solstice. This dual function of riddles, as prophetic vehicles and as diversionary games, extended throughout the medieval and Renaissance periods. Only by the middle part of the eighteenth century did they lose their mythic-divinatory value, becoming perceived almost exclusively as forms of mind-play, included as regular features in newspapers and periodicals. Famous personages also created riddles for various social-recreational reasons. In France, for example, the satirist Voltaire would regularly compose riddles for pure enjoyment or to challenge and taunt his friends and enemies.

As part of divinatory speech, riddles were seen by the Greeks to be a product of what they called *mythos*—a form of thinking based on beliefs and intuitions rather than on facts and argumentation. In contrast, they used *lógos* to designate the kind of thinking used to solve problems in geometry, to analyze facts logically, and to carry out arguments in support of an idea or theory. Aristotle was the one who coined the term *mythos* to describe the narrative plot sequences in tragedies. He defined it as the form of thinking that produced poetry, dramatic narratives, and mythic beliefs; *lógos* instead was the form of thinking that undergirded logic, philosophy, and mathematics. Socrates believed that *lógos* was innate in all human beings, teaching that individuals were born with it as part of the brain's innate cognitive capacities, and that it could thus be awakened through conscious reflection. In the *Meno*, a Socratic dialogue written by Plato (reprinted 2006), Socrates leads an untutored slave to successfully grasp a complicated geometrical problem by getting him to reflect upon the truths hidden within him through a series of questions designed to elicit his innate sense of *lógos*. The difference between the two forms of thought can be rephrased, somewhat liberally, as imagination *(mythos)* versus reasoning *(lógos)*. The main claim to be made here is that for most mathematical puzzles both modes of cognition are involved, as will be illustrated throughout this book. Indeed, the blend of imagination and reasoning that unfolds in the solution of any puzzle reveals that

Fig. 1.1 Diagram showing
that the angles of a triangle
add up to 180°
(Wikimedia Commons)

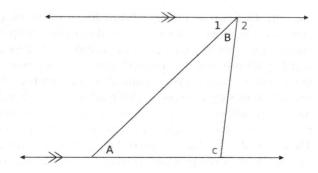

mythos and *lógos* are intrinsically intertwined and may even be structured cohesively in terms of a flow of cognition that moves from imaginative sparks (likely in the right hemisphere of the brain) to organizational and reflective thinking (in the left hemisphere).

As mentioned briefly, a problem presents information that leads directly to a solution; a puzzle, on the other hand, does not. The distinction between the two can be illustrated in the field of plane geometry. Consider a problem first:

Prove that the angles of a triangle add up to 180°.

The solution to this puzzle involves a straightforward application of known facts in a logical way, given that we have all the information needed to solve it. The relevant facts include: (a) that a straight line is equal to 180° and (b) that angles on the opposite sides of a transversal between two parallel lines are equal. So, using fact (b), in the above diagram angles 1 = A and 2 = C. Now, given (a), we can see that 1 + B + 2 = 180°. Making the appropriate substitutions, employing the axiom of equality, we get A + B + C = 180°. While there is a certain amount of background inferential thinking required here, such as realizing that (a) and (b) are essential aspects of the problem, getting the answer involves a relatively straightforward type of deductive proof that can be rehearsed and reinforced with similar problems.

Despite the straightforward method of proof used, even a simple problem such as this one shows how mathematicians turn the "meaningless marks on paper," as Hilbert characterized mathematical symbolism (cited in the preface to this book), into general principles or theorems. Since the triangle chosen was a nonspecific one, and because A, B, and C can take on any value we so desire (less than 180° of course), we have proven the proposition true for *all* triangles. This "generalization-by-demonstration" process is the sum and substance of deductive thinking. It is little wonder that the Greeks saw *lógos* as a significant feature of the human brain.

Now, consider the following, which at first glance would seem to suggest a geometry problem that can be solved just as straightforwardly as the previous one. It was devised (to the best of my knowledge) by puzzlist Martin Gardner (1994: 14):

Calculate the length of the diagonal AB in rectangle AOBC.

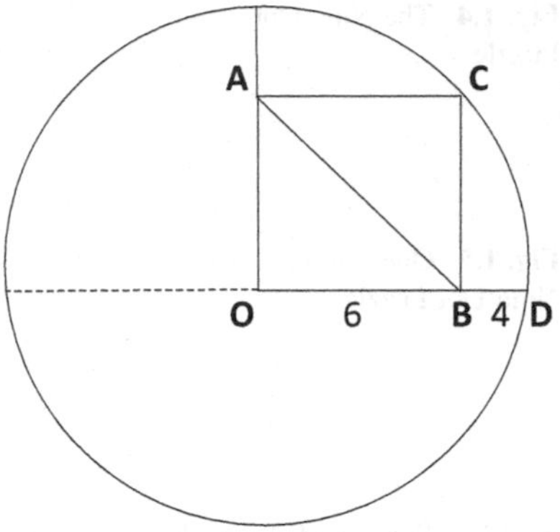

Fig. 1.2 Gardner's
geometry puzzle

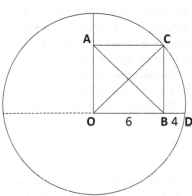

Fig. 1.3 Gardner's
geometry puzzle
re-imagined

As it turns out, it is impossible to solve it with the given information in the same way as the previous problem. For example, applying the Pythagorean theorem, suggested by the diagram, does not lead anywhere. The solution seems to be intractable. So, at this point, we must re-imagine (literally) what is known about circles and rectangles from a different perspective. Psychologically, we must shift from deduction to inference, playing around with hunches related to rectangles and circles. First, it is useful to bring some relevant information to the geometrical layout—namely, that the diagonals of a rectangle equal each other in length and that the radii of a circle are all equal to each other. This suggests drawing the other diagonal (OC) of rectangle AOBC and, of course, taking note of the fact that it is also a radius of the circle.

By doing this, we can now see that diagonal OC is not only a diagonal of the rectangle, but also a radius of the circle. Line OBD is a radius as well and it is equal to 10 (6 + 4), as shown. So, line OC, being a radius, is also equal to 10. From this, we conclude that the other diagonal of the rectangle, AB, is 10 units in length. The solution appears almost magically, rather than logically as it did in the previous problem—hence the use of the expression "Aha" to characterize the effect that such a solution tends to have on solvers (Auble, Franks, and Soraci 1979)—an effect that

Fig. 1.4 The Nine-Dot
Puzzle

Fig. 1.5 One solution to the
Nine-Dot Puzzle

will be discussed throughout this book. At this point, it is sufficient to observe that this is an apt term for describing solutions to puzzles. Incidentally, the ancient Egyptians had a similar word for this effect, as can be seen by its use in the *Ahmes Papyrus*, to be discussed below (Spalinger 1990). Clearly, the ancients unconsciously realized that there was a qualitative distinction between puzzles and problems, even though they did not distinguish lexically between the two. Problems did not produce the Aha effect; puzzles did, as can be seen by the use of the Egyptian version of "Aha" in the *Papyrus*.

The classic example used by psychologists to examine Aha thinking is the so-called Nine-Dot Puzzle, which is worth discussing briefly here:

> Without your pencil leaving the paper, can you draw four straight lines through the following nine dots? (See Fig. 1.4)

Those unfamiliar with this puzzle might attack it by joining up the dots as if they were located on the perimeter (boundary) of an imaginary square or flattened box. But this reading of the puzzle will not yield a solution, no matter how many times one tries to draw four straight lines without lifting the pencil. A dot is always left over. At this point a hunch is needed: "What would happen if the four lines were to be extended beyond the assumed imaginary box structure outline?" That hunch turns out, in fact, to be the relevant Aha insight.

One solution is shown in Fig. 1.5.

Incidentally, this puzzle is the likely source of the common expression "thinking outside the box." The reason is self-explanatory. The puzzle requires solvers to move beyond habits of mind, such as perceiving the above dots as constituting the outline of a box. The term "lateral thinking" was proposed by psychologist Edward De Bono (1970) to refer to the kind of thinking that is activated in solving the Nine-Dot Puzzle. Actually, there is an ingenious well-known version of this puzzle:

> Can you connect all the dots with only three lines? The answer is shown in Fig. 1.6.

In this case, the lines barely touch some of the dots—note that the puzzle did not ask us to go through them—thus adding a layer of ingenuity that makes it much more challenging. As far as can be determined, the first appearance of a Dot-Joining Puzzle is in the 1914 edition of puzzlist Sam Loyd's *Cyclopedia of Puzzles*. But the principle it embodies is likely older, as Martin Gardner indicates in his 1960 edition

Fig. 1.6 Different version
of the Nine-Dot Puzzle

Fig. 1.7 Loyd's Christopher Columbus' egg puzzle (Wikimedia Commons)

of Loyd's work (*The Mathematical Puzzles of Sam Loyd*). Loyd called it "Christopher Columbus' Egg" puzzle. Figure 1.7 is how he presented it in the book.

Like the development of theorems via proof, mathematicians do not stop at the singular solution that a puzzle affords; rather, they search for any generalities within the structure of the puzzle; that is, they seek to unravel what mathematical properties, if any, it might harbor. So, from lateral, imaginative thinking the mental focus shifts to logical reasoning for generalization purposes. The above puzzle is a specific, or 3 × 3 version, of a Dot-Joining Puzzle. By solving sixteen-dot, twenty-five-dot, and other *n*-dot versions, is it possible to uncover some mathematical structure hidden within it? More specifically, is there a correlation between number of dots and number of connecting lines? This mode of "post-solution thinking" is the essence of recreational mathematics, as will be discussed subsequently. But it is also characteristic of mathematical method in general. It can be called, simply, the *Generalization Principle*.

The principle can be reformulated psychologically as follows: from the creative or imaginative mode of thinking required to solve a specific puzzle (which is likely

Table 1.1 Generalizing
the Nine-Dot Puzzle

Dots	Lines required
3×3	$(3 + 1) = 4$
4×4	$(4 + 2) = 6$
5×5	$(5 + 3) = 8$
6×6	$(6 + 4) = 10$
...	...
$n \times n$	$n + (n - 2) = 2(n - 1)$

to constitute a right-hemispheric function), any hidden general principle within it can be identified by solving more complex versions (which involves the functions of the left hemisphere). In the case at hand, this entails increasing the number of dots to see what kinds of solutions the more complex versions yield. From such "complexity increments," a pattern seems to emerge through simple inspection.

Needless to say, research on this puzzle within both mathematics and psychology has revealed more complexity and implications than this (Kershaw and Ohlsson 2004). The point here is simply that it illustrates concretely how the Generalization Principle works. After the Aha insight has occurred, the reasoning part of the brain seems to kick in, figuratively speaking, to allow us to explore logically if there is any recurrent pattern that can be unpacked from the puzzle. This whole process brings to light, in microcosm, the reason that discoveries are organized into emerging new categories. Zeno's paradoxes of motion, for example, have led to the theory of limits, Alcuin's River-Crossing Puzzle prefigured modern-day critical path theory, Euler's Königsberg's Bridges Puzzle sparked the creation of graph theory and topology, and the list could go on and on. In each case, processes of generalization were involved that led mathematicians to consider implications beyond the original puzzles. As Kasner and Newman (1940: 156) aptly observed in their classic book on the mathematical imagination, the "theory of equations, of probability, the infinitesimal calculus, the theory of point sets, of topology, all have grown out of problems first expressed in puzzle form." The whole process of generalizing mathematical knowledge might be said to consist in an effort to minimize the need to repeat the type of insight thinking involved in puzzles. Once an insight is attained, it becomes useful to "routinize" or store it as a habit of mind by algorithmic generalization, so that a host of related problems can be solved as a matter of course, with little time-consuming mental effort. Such generalization is a memory-preserving and time-saving strategy. It is the rationale behind organized knowledge systems, which are essentially habitualized procedures for solving problems that would otherwise require insight thinking to be used over and over. Once such thinking has connected the dots, no pun intended, the rational part of the mind steps in to give it stability.

Although a term for "puzzle" as a distinct concept did not exist in antiquity, terms for the notion of "game" did. For instance, Archimedes' *loculus* was understood among mathematicians as an ingenious game, which bore mathematical implications. It had 14 shapes, some of which were identical. The original objective was to

Fig. 1.8 The *loculus*

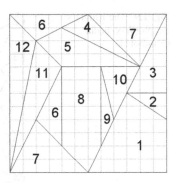

rearrange the scrambled shapes so that they always cohered into a square figure. The numerical layout in Fig 1.8 (Cutler 2003) shows two "6s" and two "7s" because the shapes they represent are identical.

A computer analysis by Bill Cutler has revealed that there are 536 distinct ways to arrange the pieces. Now, what is particularly interesting about this game is that its structure crops up subsequently in different formats and guises. For example, there are many parallels between the Archimedean *loculus* and modern-day tangram puzzles, which involve rearranging and combining seven flat geometrical shapes, called *tans* (to be discussed subsequently). In some versions of the Archimedean game, the scrambled shapes are used to form different figures (human shapes, animals, objects, etc.)—an objective that applies as well to tangrams. For the sake of historical accuracy, it should be mentioned that the *loculus* was first discussed in a treatise attributed to Archimedes called the *Ostomachion*—a work that has survived in an Arabic version and in a palimpsest of the original Greek text going back to the Byzantine era (Darling 2004: 88). So, it is not known for certain that Archimedes was the inventor of the game; he may have come across it beforehand (Netz and Noel 2007).

The distinction between puzzles and games is a blurry one. The main difference may well be that a game has an "end-state," which is announced in advance, whereas a puzzle has a nonobvious answer hidden in the statement, and which must be extracted from it. The terminology devised by semiotician Umberto Eco (1989) to refer to texts as *open* and *closed* can be applied here as well. "Closed puzzles" do not hide the answer, but give it at the start as an end-state that must be reached. The "fun" in this case is *how* to get to the end-state. Examples of closed puzzles abound. They include Sudoku, the Rubik's Cube, the Tower of Hanoi, and so on and so forth. In all closed puzzle genres, we are given a set of rules for reaching the end-state. Open puzzles, on the other hand, do not involve preannounced end-states or rules. Their answers are never obvious; the fun is figuring out what the answers are, not how to get to them via rules. Mathematical games are, by and large, closed texts. This does not imply that they do not require cleverness and adroitness of mind, but rather that we know in advance what we are aiming for, unlike open puzzles.

Many games are also interactive, involving players in competition with each other. The game of chess, for example, has an end-state (checkmate) and it involves

Fig. 1.9 The Knight's
Graph (Wikimedia
Commons)

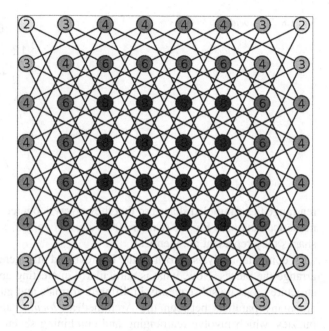

competition among people or computers in order to see who will reach the end-state first. As indicated, the distinguishing feature of all games, vis-à-vis open puzzles, is that they are rule-governed—that is, rules determine the acceptable or unacceptable actions within the game that lead to the end-state. In this book, only games used by mathematicians will be considered. Excluded are competitive recreational games such as sports.

The so-called Knight's Tour brings out the reason why games and puzzles are often indistinguishable from each other (Conrad, Hindrichs, Morsy, and Wegener 1994):

> Place a knight on a chessboard so that it visits every square once and only once.

There are a number solutions (ways of reaching the end-state), with the earliest one dating back to the ninth century in the *Kavyalankara,* a Sanskrit work on the nature of poetry, but which also includes the game in the form of a poetic figure. Figure 1.9 is the "Knight's Graph" (see Averbach and Chein 1980: 195), which shows all possible paths for the Knight's Tour—each number indicates the possible moves that can be made from that position to complete the Tour.

The first to explore the mathematical generalization properties of the Knight's Tour was Leonhard Euler, extracting from it significant insights into the nature of graphs. The Tour is an instance of a Hamiltonian graph, which will be discussed subsequently. An early heuristic for the Tour was published in 1823 by H. C. von Warnsdorf, in which the knight moves so that it always reaches the square from which it will have the fewest subsequent moves. The moves that revisit a specific square are not counted. This might result in two or more choices for which the number of subsequent moves is equal. Pohl (1967) devised an ingenious method for breaking such ties (see also Schwenk 1991). Suffice it to say that the Knight's Tour, like any classic game or puzzle, is consistent with the Generalization Principle.

The double use of the chessboard—for game-playing and for gleaning principles of mathematical structure from it—is arguably the reason why the terms *puzzle* and *game* are often used synonymously within recreational mathematics, despite the differences indicated above. This convention will be adopted in this book as well, except when a distinction is relevant. There are, for mathematical purposes, four main types of games (Dalgety and Hordern 1999): (1) movement and arrangement games, played by manipulating objects such as sticks, coins, or counters with the hands; (2) mechanical and assembly games, played by connecting pieces to make shapes; (3) board games, played on a game board; and (4) card and dice games.

As implied by the foregoing discussion, problems and puzzles (open and closed) are Q&A (Question and Answer) structures. The difference as we have seen lies in the fact that the answer in problems is a matter of straightforward analysis; while in puzzles it is not. In the case of games, there is an end-state(E), rather than, strictly speaking, an answer (A) and a set of rules (R) for reaching it:

Problem
Q → A (the question leads directly to an answer)

Puzzle
Q → (A) (the answer to the question is not immediately obvious)

Game
Q → R → E (the end-state is reached via application of specific rules to the initial state)

The Q&A structure is not limited to problems and puzzles, of course. The Socratic dialogue and dialectic reasoning generally exhibit the same frame of mind. The term *dialectic* is attributed to the pre-Socratic philosopher Zeno of Elea by Aristotle, but it is implicit in Plato's dialogues, which were constructed to illustrate Socrates' own suggestive dialectical method. Mysteries (real and fictional) also have a general dialectic structure, since they present information that hides an answer. When it comes, as in a mystery story, we feel relief; when it does not, as in some real mystery, then we continue to search for an answer. Essentially, the dialectic mind aims to probe the nature of things by posing questions about them, and then seeking answers to them wherever and whenever possible. Every new and challenging puzzle engages us in an internal dialectic with its contents, since it impels us to find an answer. When an answer is not forthcoming, we are left in a frustrating and even anguished state of mind.

In all classic math puzzles, the starting point in the dialectic process is to develop a hunch in view of obtaining an answer that makes sense under the circumstances. The American mathematician and philosopher, Charles Sanders Peirce, saw hunches as the sparks for solving puzzles and for grasping new ideas associated with them, after which reasoning takes over to complete the mental process. He called this initial form of thinking "abduction" (Peirce 1931–1958, volume V: 180):

> The abductive suggestion comes to us like a flash. It is an act of insight, although of extremely fallible insight. It is true that the different elements of the hypothesis were in our minds before; but it is the idea of putting together what we had never before dreamed of putting together which flashes the new suggestion before our contemplation.

Hunches are the brain's attempts to understand what something might mean or imply initially by considering images and associations beyond the obvious or the habitual, as we saw with the Nine-Dot Puzzle. These eventually lead to inferences or possibilities (sometimes even hypotheses) through a series of inner visualizations. These might then generate the required Aha insight. When they do, a solution comes spontaneously into mental focus. Of course, other cognitive processes, modalities, or phases might be involved, including deduction, induction, trial and error, and so on. But by and large abduction is sufficient to explain the initial stages of puzzle-solving, open and closed. Hunches can occur through analogies, inferences based on previous experiences, and other contextual factors. This is arguably why puzzles are universal products of human ingenuity, independent of historical time and cultural spaces. As Henry Dudeney (1958: 12) aptly put it:

> The curious propensity for propounding puzzles is not peculiar to any race or any period of history. It is simply innate … though it is always showing itself in different forms; whether the individual be a Sphinx of Egypt, a Samson of Hebrew lore, an Indian fakir, a Chinese philosopher, a mahatma of Tibet, or a European mathematician makes little difference.

The *Ahmes Papyrus*

Because of the practical importance of counting and measuring, mathematics has always been considered to constitute part of a basic educational curriculum, alongside literacy and reading. One of the oldest math texts, dating to before 1650 BCE, is an Egyptian papyrus that was likely used as a school textbook. Eighteen and a half feet long and thirteen inches wide, the work is called either the *Ahmes Papyrus*, after the Egyptian scribe who copied it, or the *Rhind Papyrus*, after the Scottish lawyer and antiquarian, A. Henry Rhind, who purchased it in 1858 while vacationing in Egypt. In addition to over eighty challenging math problems (most of which are de facto puzzles), the *Papyrus* contains tables for the calculation of areas, the conversion of fractions, the structure of elementary sequences, and extensive information about measurement. The earliest known symbols for addition, subtraction, and equality are also found in this truly remarkable work (Gillings 1972: 246–247). Figure 1.10 is a page from the *Papyrus*.

As the scribe Ahmes relates in his introductory remarks, the work is his copy of an older work by an anonymous author (Chase 1979: 27):

> This book was copied in the year 33, in the fourth month of the inundation season, under the majesty of the king of Upper and Lower Egypt, 'A-user-Re,' endowed with life, in likeness to writings made in the time of the king of Upper and Lower Egypt, 'Ne-ma'et-Re'. It is the scribe A'h-mose who copies this writing.

Fig. 1.10 Page from the *Ahmes Papyrus*

The original work was thus written during the same era in which another famous manuscript of Egyptian mathematics, the *Moscow Papyrus* (named after its current location), was composed. The king A-user-Re has been identified as a member of the Hyskos dynasty, which governed Egypt around 1650 BCE. The king Ne-ma'et-Re was Amenemhet III, who reigned from 1849 to 1801 BCE—the period when the original was thus written.

The *Papyrus* is divided into three main parts—problems in arithmetic and algebra, problems in geometry, and miscellaneous problems. This division suggests that it may have been conceived both as a treatise in early mathematical classification and a school textbook. This is not unusual since, in antiquity, the two domains were not seen as necessarily distinct. Euclid's *Elements*, for example, was intended as a teaching textbook, at the same time that it systematized the methods of proof. The *Papyrus* was translated into German in 1877 and into English in 1923. The first extensive edition of the work was completed in 1929 by Arnold Buffum Chase—an edition that made the text accessible for the first time to a large public (Chase 1979). It is now preserved in the permanent collection of the British Museum, which came into its possession after Rhind's death (see Peet 1923, Gillings 1961, 1962, 1972, Chase 1979, Robins and Shute 1987, Olivastro 1993: 31–64).

Many of the "problems" in the *Papyrus* were patently "puzzles," as defined here. As Petkovic (2009: 2) also states, the work reveals "that the early Egyptians based their mathematics problems in puzzle form." Solving Problem 56, for instance, leads to an insight that was original for the era in which the *Papyrus* was written—namely, that the height of a pyramid can be related to the size and slope of each of its triangular walls. It is labeled a "problem" in all translations of the text, but upon closer scrutiny it can be seen to have the features of a puzzle, given that the answer was not a simple calculable one, but rather requires considerable imaginative thinking for the era in which it was written.

Consider Problem 79, which has become widely known. It presents information in the form of an inventory without an appurtenant question, implying that the solver had to figure out what hidden pattern was involved:

Houses	7
Cats	49
Mice	343
Sheaves of wheat	2401
Hekats of grain	16,807
Estate	19,607

The solution seems intractable at first. By looking closely at the numbers representing the items of the estate (*houses, cats,* etc.), an Aha insight crystallizes—the first five numbers are successive powers of 7: $7 = 7^1, 49 = 7^2, 343 = 7^3, 2401 = 7^4$, and $16,807 = 7^5$. But this pattern does not apply to the last figure, 19,607. To discern the relevance of this number, we need another hunch. Since the first five numbers represent items that are found in an estate, then the last number, 19,607, suggests the sum of the items, especially given the fact that it is placed at the end of the itemization, like the sum in many problems in arithmetic. And this is indeed the final touch to the solution: $7 + 49 + 343 + 2401 + 16,807 = 19,607$.

It should be noted that in his translation of the *Papyrus*, Peet (1923) reformulates the puzzle statement as follows—a reformulation that diminishes the challenge involved in solving it, transforming it into a problem:

> Seven houses: in each are 7 cats; each cat kills 7 mice; each mouse would have eaten 7 ears of spelt; each ear of spelt will produce 7 hekat. What is the total of all of them?

The significance of this puzzle does not stop at the solution. It conceals what can be called a "puzzle archetype," to use psychologist Carl Jung's (1983) notion in a liberal fashion throughout this book. This can be defined as a particular mathematical motif, pattern, or idea that surfaces in other puzzles even when there are no historically established connections among them. The archetype in Problem 79 is based, arguably, on the mystical meanings that the number seven had in antiquity, finding expression in a simple puzzle format. The unconscious fascination with this number may be the reason why the same archetype surfaces in Leonardo Fibonacci's *Liber Abaci* of 1202. In this case, the Italian mathematician simply added another power of 7, namely 7^6, to the puzzle (Fibonacci 2002):

> Seven old women are on the road to Rome. Each woman has seven mules, each mule carries seven sacks, each sack contains seven loaves, to slice each loaf there are seven knives, and for each knife there are seven sheaths to hold it. How many are there altogether, women, mules, sacks, loaves, knives, sheaths?

Fibonacci could not possibly have known about Ahmes' puzzle, because the existence of the *Papyrus* was not known at the time, nor had hieroglyphic-hieratic writing been deciphered. But the similarity between the two puzzles is unmistakable. Interestingly, in eighteenth-century England a version of the same archetype appeared clothed as a popular nursery rhyme:

As I was going to St. Ives
I met a man with seven wives.
Each wife had seven sacks,
Each sack had seven cats,
Each cat had seven kits.
Kits, cats, sacks, wives,
How many were going to St. Ives?

That version, however, contained a clever trap. The anonymous puzzlist asked how many kits, cats, sacks, and wives were *going to* St. Ives, not *coming from* it. This means that only one person was *going to* St. Ives—the narrator of the rhyme. All the others were, by implication, making their way *out of* the city. As these historically unrelated coincidences with Ahmes' puzzle show, archetypes surface in different time periods and in different formats. Of course, the use of 7 by Ahmes may have a more mundane explanation, as Gillings (1972: 168) suggests: "The number 7 often presents itself in Egyptian multiplication because, by regular doubling, the first three multipliers are *always* 1, 2, 4, which add to 7." But, as Maor (1998: 13) remarks, this explanation is unconvincing because "it would equally apply . . . in fact to all integers of the form 2^n-1."

The *Papyrus* starts off with the following miniature poetic statement:

Accurate reckoning,
the entrance into the knowledge of all existing things and all obscure secrets.

This suggests that Ahmes (or the original author) understood that puzzles were experiments in discovery, since a puzzle illuminates some hidden pattern (an obscure secret). Even the *Papyrus's* title—*Directions for Attaining Knowledge of All Dark Things*—alludes figuratively to the fact that puzzles allow mathematics to uncover hidden or inherent structure. This can be called "Ahmes' legacy." Consider, as another case-in-point, Ahmes' estimation of the value of π. The relevant puzzle is Problem 48:

What is the area of a circle inscribed in a square that is 9 units on its side?

Fig. 1.11 Problem 48 from
the *Ahmes Papyrus*

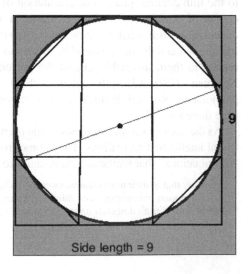

The clever Ahmes (or the original author) solved it with the following Aha insight: "What if the circle is transformed into a polygon?" He proceeded to do exactly that by trisecting each side of the square, as shown in the diagram, in order to produce nine smaller squares within it (each 3×3). He then drew the diagonals in the corner squares. Such modifications to the diagram produced an octagon, which Ahmes assumed to be close enough in area to the circle for the purposes of his puzzle. The area of the octagon is equal to the sum of the areas of the five inner squares (which form the outline of a cross) plus half the sum of the areas of the four corner squares (= the sum of two squares). The octagon's area is thus equal to the sum of the areas of seven small squares. The area of one square is, of course, 3×3, or 9 square units. The total area of seven such squares is, therefore, 9×7, or 63 square units. With a bit of convenient cheating, the resourceful Ahmes assumed that the circle's area was virtually equal to 8^2. He then estimated the value of π as follows (using modern notation):

Area of circle:	$= \pi r^2 = 64$
Diameter:	$= 9$
Radius (r):	$= 9/2$
So, r^2	$= (9/2)^2 = 20.25$
Thus, since πr^2	$= 64$
and r^2	$= 20.25$
π	$= 64/20.25$
π	$= 3.16049\ldots$

This same method of computing π surfaces in the mathematics of various other periods and in different cultural contexts, thus suggesting the unconscious operation of an archetype, which can be called "polygoning the circle" (a version of "squaring the circle"). As is well known, it crystallized over 1000 years later in Archimedes' famous demonstration. Archimedes inscribed a polygon with 96 sides in a circle to calculate the value of π as somewhere between 3 and 10/71 and 3 and 1/7. In 264 CE, the Chinese mathematician Liu Hui inscribed a polygon of 3072 sides to calculate π to the fifth decimal place. The calculation of π has not been a trifling matter in the course of human civilization. A world in which π is not known is, of course, conceivable. But what we now know about certain natural objects in the world, like the sun and the tides, would be much more rudimentary, since equations that are related to them invariably enfold the presence of π within them. As Kasner and Newman (1940: 89) aptly put it, without π "our ability to describe all natural phenomena, physical, biological, chemical or statistical, would be reduced to primitive dimensions."

As the above example suggests, some (perhaps many) puzzles are hardly ever just trivial intellectual recreations. They come from what Dehaene (1997: 151) calls the "illuminations" that mathematicians claim to see within their minds:

> They say that in their most creative moments, which some describe as "illuminations," they do not reason voluntarily, nor think in words, nor perform long formal calculations. Mathematical truth descends upon them, sometimes even during sleep.

The term "illumination" is quite appropriate, as Ahmes intuitively realized by pointing out that the kind of "accurate reckoning" required to unravel "obscure secrets" made them "visible," which is another way of saying "illuminated." Once the secret is out, the Generalization Principle becomes operative, spreading the secret in a manner that is analogous to how "internet memes" spread in cyberspace today. The term *meme* was introduced by Richard Dawkins (1976) to refer to an element of a culture or system of behavior that is passed from one individual to another by nongenetic means. This notion will be discussed critically in due course. For now, it is a convenient one for describing how ideas discovered via puzzles then spread among mathematicians—no more, no less. Above all else, the *Ahmes Papyrus* is an "archeological mathematical treasure," for a systematic study of the earliest forms and functions of math puzzles and the various archetypes they enfold.

Puzzles in Mathematics

Historians of mathematics trace its origins as a distinct or autonomous discipline to the notion of *proof*, which was given its first explicit formulation around 500s BCE Greece. The formal discussion and elaboration of this notion was accomplished brilliantly by Euclid in his *Elements* around 350 BCE—a treatise that (as mentioned) was a founding text for mathematics as a theoretical enterprise and a textbook for teaching it (Euclid 1956). Euclid established general theorems from specific cases (or propositions) through various methods of proof, devising some of them imaginatively himself, that became cornerstones of mathematics and remain so to this day.

A classic example of what proof can accomplish is Pythagoras's right-triangle theorem, which showed that the sides of a right triangle have the same abstract relation to each other, no matter the size of the triangle. Practical knowledge of the structural features of right triangles was common in many parts of the ancient world (from China and India to Africa and the Middle East) before Pythagoras (Strohmeier and Westbrook 1999). Around 2000 BCE, the Egyptians discovered practically that knotting and stretching a rope into a triangle making sides of 3, 4, and 5 units in length produced a right triangle, with 5 the longest side. Pythagoras's theorem proved that this specific case was really only one instance of a more general pattern—if the square of the length of the hypotenuse equals the sum of the squares of the lengths of the other two sides, then a right triangle is always formed. Knotting any three stretches of rope according to this general theorem—for example, 6, 8, and 10 units—will produce a right triangle because $6^2 + 8^2 = 10^2$ ($36 + 64 = 100$). In effect, the theorem revealed an "obscure secret" about the world, to cite Ahmes. As Jacob Bronowski (1973: 168) has aptly written, it was a secret that harbored within it one of the laws of the universe:

> The theorem of Pythagoras remains the most important single theorem in the whole of mathematics. That seems a bold and extraordinary thing to say, yet it is not extravagant; because what Pythagoras established is a fundamental characterization of the space in which we move, and it is the first time that it is translated into numbers. And the exact fit of the numbers describes the exact laws that bind the universe. If space had a different symmetry the theorem would not be true.

Now, it can be argued that proofs, as elegant and ingenious as they may be, are not always the sparks for generating or discovering new math ideas, even though they have at times been critical in this domain (see Benson 1999). And even when proofs lead to some discovery or some verification, the spark comes first from an abduction that triggers the relevant direction of thinking. An early classic example is, actually, Euclid's proof that the prime numbers are infinite, since the initial "illumination" for the proof is an abduction.

Euclid knew what Pythagoras had shown previously, namely that we can divide the integers into those that can be decomposed into factors—composite numbers— and those that cannot. The numbers 12, 42, and 169, for instance, are all composite because they are the products of smaller factors: $12 = 2 \times 2 \times 3$, $42 = 7 \times 2 \times 3$, $169 = 13 \times 13$. The factors themselves were designated prime numbers. The first nine primes are: 2, 3, 5, 7, 11, 13, 17, 19, and 23. Now, even a cursory examination of the integers laid out in order on a number line will reveal that there are fewer and fewer primes as the numbers increase along the line to infinity: 25% of the numbers between 1 and 100, 17% of the numbers between 1 and 1000, and 7% of the numbers between 1 and 1,000,000 are primes. Thus, it appears logical to assume that the primes must come to an end at some point. Common sense would also have it that if a number is big enough, it must be the product of other, smaller numbers. But, with a veritable Aha abduction, Euclid proved that this is not so.

The method of proof that Euclid used, as is well known, is called *reductio ad absurdum*, literally "reduced to the absurd." It was devised originally by Zeno of Elea. As Berlinski (2013: 83) points out, this kind of proof is truly ingenious because it "assigns to one half [of the mind] the position one wishes to rebut, and to the other half, the ensuing right of ridicule." A paraphrase of the proof based on a standard (and contemporary) version will be used here. It nonetheless captures the essence of Euclid's ingenious abduction. He started with the contrary hypothesis—namely, that there is a finite set of primes (P):

$$P = \{p_1, p_2, p_3, \ldots p_n\}$$

The symbol p_n stands for the last (and largest) prime; each of the other symbols stands for a specific prime: $p_1 = 2$, $p_2 = 3$, and so on. Next, he multiplied all the primes in the set, in order to produce a composite number, C, that is divisible by any of the primes in P:

$$C = \{p_1 \times p_2 \times p_3 \times \ldots \times p_n\}$$

At this point, Euclid had his Aha moment: What would happen if we added 1 to C: $C + 1 = \{p_1 \times p_2 \times p_3 \times \ldots \times p_n\} + 1$. Let's call the number produced in this way M, rather than $C + 1$:

$$M = \{p_1 \times p_2 \times p_3 \times \ldots \times p_n\} + 1$$

Clearly, M is not divisible by any of the primes in P, because a remainder of 1 would always be left over. So, the number M is either: (a) a prime number that is

Fig. 1.12 The Sieve of
Eratosthenes

1	2	3	4	5	6	7	8	9	10
11	12	13	14	15	16	17	18	19	20
21	22	23	24	25	26	27	28	29	30
31	32	33	34	35	36	37	38	39	40
41	42	43	44	45	46	47	48	49	50
51	52	53	54	55	56	57	58	59	60
61	62	63	64	65	66	67	68	69	70
71	72	73	74	75	76	77	78	79	80
81	82	83	84	85	86	87	88	89	90
91	92	93	94	95	96	97	98	99	100

not in P and thus greater than p_n, or (b) a composite number with a prime factor that, as just argued, cannot be found either in the set P and thus also greater than p_n. Either way, there must always be a prime number greater than p_n. In conclusion, the prime numbers are infinite. The Aha insight of adding 1 to C might seem trivial, but this is so only after it has been envisioned by Euclid's imagination and presented to us accordingly. It is one of those illuminations of which Dehaene speaks.

As happens throughout the history of mathematics, a solution or proof such as this one contains many further dialectic ideas to be unpacked from it. How can we tell if a number is prime? How many primes are there? Is there a pattern to the occurrence of the sequence of primes? One of the first questions that mathematicians asked, after Euclid's proof of the infinity of primes, is the following one: Is there a rule or algorithm that will generate primes and only primes? One of the first to put forward such an algorithm was Eratosthenes in the 200s BCE. He constructed a ten-by-ten square listing the first 100 numbers, known as the *Sieve of Eratosthenes* (Fig 1.12).

The first prime number in the sieve is 2. So, we cross out every second number after 2. This eliminates all numbers that can be divided evenly by 2, except for 2 itself. Then, we cross out every third number after 3, since these are multiples of 3. This step eliminates all the numbers that can be divided evenly by 3, except for 3 itself. Then, since 4 is already crossed out, we move on to 5. We cross out every fifth number after 5, and so on. Numbers that are not crossed out can be thought of as having passed through a sieve (strainer) that has caught all the primes—hence this characterization of the diagram. Eratosthenes' sieve ends up "trapping" 25 primes. Needless to say, it would be a gargantuan task to set up a sieve to trap the primes even among the first 1000 numbers. So, it really is not a practicable algorithm, but it shows ingenuity nonetheless. With the aid of modern computers, the sieve has been revisited, expanding its trapping power considerably. But the sieve does not reveal an underlying rule for generating primes.

One of the most intriguing and difficult dialectic outcomes related to the primes is known as the Riemann Hypothesis (Derbyshire 2004, Du Sautoy 2004, Sabbagh 2004, Wells 2005, Rockmore 2005). In 1859, Bernhard Riemann presented a paper to the Berlin Academy titled "On the Number of Prime Numbers Less Than a Given Quantity" in which he put forth a hypothesis that remains unsolved to this day. Riemann never provided a proof and his housekeeper burnt all his personal papers after his death. It is a proof that is waiting to be made, even though its exploration has already led to several significant discoveries in mathematics (Erdös 1934). Riemann argued that the thinning out of primes involves an infinite number of "dips" called "zeroes," on the line, and it is these zeroes that encode all the information needed for testing primality. So far no vagrant zero has been found, but at the same time no convincing and accepted proof of the hypothesis has come forward.

From previous work, Riemann knew that the number of primes around a given number on the line, n, equals the reciprocal of the natural logarithm of that number—the number of times we have to multiply e ($= 2.71828$) by itself to get the number. Riemann showed that at around one million, whose natural logarithm is about 3, every 13th number or so is prime. At one billion, whose natural logarithm is 21, about every 21st number is prime. Riemann asked why primes were related to natural logarithms in this way. In sum, if the hypothesis is right, then we will know how the primes thin out along the number line. So far computers have been able to verify the hypothesis for the first 50 billion. What kind of proof would be involved in showing that it applies to all? Incredibly, the zeta function (as it is called technically) is related to the energies of particles in atomic nuclei, to aspects of the theory of relativity, and other natural phenomena.

Pythagoras believed that the divinities had allowed him, a mere mortal, to catch a unique glimpse into the *raison d'être* of one of the secret laws governing the cosmos with his theorem. As indicated, the $c^2 = a^2 + b^2$ relation was known as a geometrical pattern in various parts of the world, suggesting that it too harbored an archetype based on the structure of triangles. One of the earliest demonstrations of its validity appeared around 600 years before Pythagoras, and in a totally different part of the ancient world—in *The Arithmetic Classic of the Gnomon and the Circular Paths of Heaven*, a Chinese treatise that is dated to around 1100 BCE (Li and Du 1987). It is also proved in the *Nine Chapters on the Mathematical Art*, a Chinese collection of mathematical ideas published in the third century BCE, although it goes back further in time (Swetz and Kao 1977).

The Pythagorean theorem turned practical and intuitive knowledge into theoretical epistemic knowledge. It was not just a recipe of how to construct right triangles; it was the brain's way of revealing something recurrent in the universe. The Pythagorean triples that can be derived from $c^2 = a^2 + b^2$ even transcend the specific properties of right triangles, having become vehicles of study within number theory. One offshoot is Fermat's Last Theorem of 1637:

> If an integer n is greater than 2, then the equation $c^n = a^n + b^n$ has no solutions in nonzero integers a, b, c.

As is well known, Fermat wrote in the margin of his copy of Diophantus' *Arithmetica* that there is no whole-number solution of $c^n = a^n + b^n$ if n is greater than 2. He wrote in the same margin that he had found a straightforward proof of this fact, but that there was not enough room to write it down. Fermat never published his purported proof and no proof was found for more than 350 years after. In 1993, British mathematician Andrew Wiles announced that he had proved the theorem. Wiles published his complete proof, with certain corrections, in 1995. However, it is clearly not the kind of proof that Fermat had envisioned, since the mathematics involved came long after Fermat. So, there are two possibilities: Fermat may have believed erroneously that he had a proof; or the kind of proof that he envisioned is still an "obscure secret" that cries out for discovery. To this day, Fermat's Last Theorem continues to engender debate and discussions. It is, in a fundamental way, a veritable puzzle, whose answer remains elusive—at least in the way Fermat had envisioned it. We shall return to the theorem in the final chapter.

The whole notion of proof as the foundation of mathematics needs some further elaboration. The ancient mathematicians transformed hunches and guesses into formats that allowed them to explore quantitative and spatial structures. A proof such as the Pythagorean theorem packs a considerable amount of hidden information within it, which crystallizes only after the relevant demonstration is found. This pattern has permitted mathematicians to let go physically of their environment in order to grasp the world in abstract ways and then to act upon it in concrete ways. Stewart (2008: 46) summarizes this incredible feature of the mathematical mind as follows:

> Using geometry as a tool, the Greeks understood the size and shape of our planet, its relation to the Sun and the Moon, even the complex motions of the remainder of the solar system. They used geometry to dig long tunnels from both ends, meeting in the middle, which cut construction time in half. They built gigantic and powerful machines, based on simple principles like the law of the lever.

This whole line of argument leads to a philosophical reconsideration of the Platonic-versus-constructivist view of mathematics. Do we discover mathematics or do we invent it and then discover that it works? Plato believed that mathematical ideas pre-existed in the world and that we come across them or, perhaps, extract them from the world through *lógos*. Just like the sculptor takes a clump of marble or clay and gives it the form of a human body, so too mathematicians take a clump of reality and give it symbolic (numerical or geometrical) form. In both representations we discover many more things about the body and about mathematics respectively. The truth is already in the amorphous clump; it takes the mathematician to give it a shape. Some mathematicians find this perspective difficult to accept, leaning toward constructivism, or the idea that mathematical objects are constructed and tell us what we want to know about the world. But, as Berlinski (2013: 13) suggests, the Platonic view is not so easily dismissible:

> If the Platonic forms are difficult to accept, they are impossible to avoid. There is no escaping them. Mathematicians often draw a distinction between concrete and abstract models of Euclidean geometry. In the abstract models of Euclidean geometry, shapes enjoy a pure Platonic existence. The concrete models are in the physical world.

Without the Platonic models, the concrete ones would generate very little (if any) interest. Moreover, there might be a neurological basis to Platonism. As neuroscientist Pierre Changeux (2013: 13) muses, Plato's trinity of the Good (the aspects of reality that serve human needs), the True (what reality is), and the Beautiful (the aspects of reality that we see as pleasing) is actually consistent with the notions of modern-day neuroscience:

So, we shall take a neurobiological approach to our discussion of the three universal questions of the natural world, as defined by Plato and by Socrates through him in his *Dialogues*. He saw the Good, the True, and the Beautiful as independent, celestial essences of Ideas, but so intertwined as to be inseparable...within the characteristic features of the human brain's neuronal organization.

But what this line of argumentation ignores is that what we perceive and portray mathematically as reality may be constrained by the limits of our mind and, thus, of our own making. Moreover, Plato's view would mean that we never should find faults within our systems of mathematics and logic, for then it would imply that the brain is faulty as a discoverer of reality or that reality itself is faulty. As it turns out, this is what Gödel's (1931) undecidability theorem implied (as will be discussed subsequently). But then, if mathematics is a human invention, why does it lead to demonstrable discoveries, both within and outside of itself? René Thom (1975) referred to discoveries in mathematics as "catastrophes" in the sense of cognitive events that subvert or overturn existing knowledge (Wildgen and Brandt 2010). Thom names the process of discovery "semiogenesis," defining it as the emergence of "pregnant" forms within symbol systems themselves. These emerge by happenstance through contemplation and manipulation of the forms. As this goes on, every once in a while, a cognitive catastrophe occurs that leads to new insights, disrupting previous knowledge. Now, while this provides a description of what happens—discovery is indeed catastrophic—it does not tell us why the brain produces catastrophes in the first place. Perhaps the connection between the brain, the mind, and the world will always remain a mystery. Actually, Thom's ideas might characterize how discoveries emerge from the classic puzzles, which can, in fact, be characterized as cognitive catastrophes that upturn an existing system or else refashion it in new and ingenious ways.

Thom's theory raises several other questions, such as: Does catastrophic discovery exist as a cross-species ability? Actually, this question was entertained by the Alexandrian geometer Pappus, while contemplating a problem: What is the most efficient way to tile a floor? There are three ways to do so with regular polygons (Flood and Wilson 2011: 36)—with equilateral triangles, equal four-sided figures, or regular hexagons. The hexagonal pattern has the most area coverage. Incredibly, bees instinctively use this pattern to build their honeycombs. In other words, bees choose the pattern with the most angles, because it holds more honey than the other ones. This is a truly mind-boggling fact, raising a whole series of questions that cannot be taken into account here. As the Estonian biologist Jakob von Uexküll (1909) might have put it, the internal modeling system of bees (which he called the *Innenwelt*) is well adapted to understanding the external world (the *Umwelt*) that they inhabit, producing instinctual models of that world. Mysteriously, some of these models overlap with human models, suggesting that there is something beyond species-specificity that unites animals and humans alike. Interestingly, Kepler made the following relevant observation (cited in Banks 1999: 19):

What purpose had God in putting these canons of architecture into the bees? Three possibilities can be imagined. The hexagon is the roomiest of the three plane-filling figures (triangle, square, hexagon); the hexagon best suits the tender bodies of the bees; also labour is saved in making walls which are shared by two; labour would be wasted in making circular cells with gaps.

Hexagonal structure also occurs in the molecular configuration of snowflakes and ice crystals. It is little wonder, as Banks (1999: 19) puts it, "that currently mathematicians and scientists are devoting much more attention to research on topics advocated by Kepler." Perhaps all this shows is that the brain is an organ that is tuned into the structure of the world (since it arises from it) and then provides glimpses of that structure a little at a time in its own way. It does so frequently by producing discoveries through dialectic reasoning.

The proof that $\pi = 3.16$ was the result of an Aha insight, as argued. Many of the problems in the *Ahmes Papyrus* involve similar thinking. Consider Problem 24:

A quantity and its 1/7 added together become 19. What is the quantity?

Ahmes' puzzle implies using a linear equation—an insight that was truly remarkable for the era in which the puzzle was formulated. So, using modern notation, if $x = quantity$, the relevant equation is shown below:

$$x + 1/7x = 19$$

The answer is $133/8 = 16.625$. Looking at this with modern eyes it can be considered to be a straightforward problem. But with ancient Egyptian eyes this would have constituted a veritable puzzle that required quite a bit of imaginative thinking.

It was the connections made by Pythagoras between geometrical forms and numbers that led to early discoveries in number theory. One of these was that square numbers can be displayed geometrically.

This was, at one level, a simple observation, but by connecting geometrical structure with numbers a hidden pattern manifested itself—each square number is the sum of consecutive odd integers (the symbol $[2n - 1]$ stands for any odd number):

$1 = 1$
$4 = 1 + 3$
$9 = 1 + 3 + 5$
$16 = 1 + 3 + 5 + 7$
$25 = 1 + 3 + 5 + 7 + 9$
...
$n^2 = 1 + 3 + 5 + 7 + 9 + \ldots + (2n - 1)$

The pattern can now be explained in more abstract terms—to form each new square figure a successive odd number of squares to the preceding figure must be

Fig. 1.13 Pythagorean square numbers

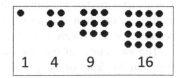

added. By connecting numbers with shapes, Pythagoras united arithmetic with geometry. This amalgam became the basis of mathematics as a theoretical mode of inquiry.

The unconscious sense that puzzles are discovery devices is Ahmes' legacy, as suggested above. But mathematicians have also used puzzles to outwit each other playfully, indicating that the brain is both a discovery organ and a ludic organ that underlies playful trickery (Hellman 2006). There is a legendary dispute between Girolamo Cardano and Niccolò Tartaglia over the authorship of the solution to cubic equations. The quarrel attracted many people, mathematicians and non-mathematicians alike, being characterized by shifting alliances, betrayals, and other forms of human mischief. Whether or not the feuds occurred in this way, the point is that they constituted "puzzle challenges" between two brilliant mathematicians, bringing out the growing awareness that puzzles had great import as well as providing recreational ludic contexts, much like riddles do.

Puzzle collections are found throughout the medieval period. The *Greek Anthology*, for example, was a compilation of literary verses, epigrams, riddles, and mathematical puzzles, known far and wide and likely used for the education of privileged children. Some credit the poet Metrodorus (c. 500 CE) with writing it, although the author is not known for certain. The *Anthology* is interesting historically because it contains many of the same puzzle archetypes found in the Ahmes Papyrus. It is useful here to discuss one of its puzzles (Wells 1992: 23):

> I desire my two sons to receive the thousand staters of which I am possessed, but let the fifth part of the legitimate one's share exceed by ten the fourth part of what falls to the illegitimate one.

Using modern algebraic notation, this puzzle can be solved easily, making it a problem as defined above, but one cannot help but wonder how people in the sixth century CE went about solving it. In that era it required, arguably, a large dose of ingenuity to solve, since there is no evidence that such puzzles were solved with algebraic techniques. A contemporary solution can be elaborated easily as follows. We start by letting x stand for the amount of staters given to the legitimate son. From this, it follows that $(1000-x)$ represents what was given to the illegitimate son. Now, the puzzle states that the fifth part of the legitimate son's share, or $1/5x$, will exceed the fourth part of the illegitimate son's share, or $1/4 (1000-x)$, by 10 staters. This translates into the following equation:

$$1/5x - 1/4 (1000 - x) = 10$$

The solution is $x = 577$ and 7/9 staters. This is what the legitimate son received; the other son thus got (1000-577 and 7/9), or 422 and 2/9 staters. Incidentally, as is the case with other puzzles from antiquity and the medieval period, the statement reveals an aspect of the cultural biases of the era in which the puzzle was devised— legitimate children are more meritorious of inheritance than illegitimate ones, because the latter were born out of wedlock.

The awareness of the importance of puzzles to impart math knowledge and to stimulate the kind of imaginative thinking that is involved in doing mathematics is implicit in an anthology of over fifty puzzles composed by the medieval scholar, Alcuin of York, who was an advisor to the emperor Charlemagne. It is titled *Propositiones ad acuendos juvenes* (translated into English as "Problems to Sharpen the Young" by Hadley and Singmaster in 1992). There is some question as to whether Alcuin was the actual inventor or the compiler of the puzzles, translating them into Latin. Whatever the case, the likely objective of the book, as implied by its title, was to train medieval youth in basic mathematical thinking. Below is one of the clever puzzles—Puzzle 34—that the anthology contains:

> A certain head of a household had 100 servants. He ordered that they be given 100 modia of corn as follows. The men should receive three modia; the women, two; and the children, half a modium. Thus how many men, women, and children were there?

Needless to say, this is an example of a Diophantine problem in which there are more variables than there are equations. The number of solutions is thus theoretically infinite, but its unique solution can be wrested out of the two equations on the basis of the given facts. Letting m, w, and c stand for the number of men, women, and children, respectively, the statement of the puzzle can be converted into the following two equations, again using modern-day notation. For the sake of illustration, it is worthwhile going through the solution here in detail, since it shows how the mathematical mind proceeds from an insight into a stepwise method of analysis, that is, from imagination to ratiocination:

$$m + w + c = 100 \tag{1}$$

$$3m + 2w + 1/2c = 100 \tag{2}$$

Equation (1) states in algebraic terms that there are 100 people in all, and Eq. (2) represents the ways in which the 100 modia are distributed, that is, "m" men will receive "$3m$" modia in total, "w" women "$2w$" modia in total, and "c" children "$1/2c$" modia in total, for a grand total of 100 modia. Only positive integral values of m, w, and c are meaningful here, since fractional or negative integral values would have no real-life sense—people cannot be split into fractions, nor can negative numbers represent them. That is the required Aha insight, which, as Alcuin must have understood, cannot be taken for granted in students. Thus c, which stands for the number of children, must be divisible by 2, otherwise $1/2c$ in the second equation would not yield an integer. Algebraically, this can be expressed by replacing c with $2n$, the general form of an even integer (a form that reflects the fact that any number, n, when multiplied by 2 will always yield an even number). Substituting $c = 2n$ into the equations above produces the following:

$$m + w + 2n = 100 \tag{3}$$

$$3m + 2w + n = 100 \tag{4}$$

Now, we can multiply Eq. (4) by 2, yielding the following equivalent equation:

$$6\,m + 4w + 2n = 200 \tag{5}$$

We can then subtract Eq. (3) from Eq. (5), an operation that reduces the problem to a single equation:

$$5\,m + 3w = 100 \tag{6}$$

From this, it can be seen that:

$$w = (100 - 5\,m)/3 \tag{7}$$

We note that m, which stands for the number of men, must be less than 99, because, if it were assigned a value of 99 or 100, the total number of people (including women and children) would be greater than 100—another important insight that cannot be taken for granted. So, m must be a positive integral value less than 99 which, when substituted into Eq. (7), will produce a positive integral value for w. If we assign the value 1 to m, a fractional value for w will result. If we let $m = 2$, however, then the value of w turns out to be 30. This is a definite possibility to consider further. So, we can go back to one of the two equations above, say, (3) $m + w + 2n = 100$, and substitute $m = 2$ and $w = 30$ in it. From this, the value of n turns out to be 34. Now, since $2n = c$, it is obvious that $c = 68$. We now have the solution, since 2 men, 30 women, and 68 children add up to 100 people in total. To check that our solution is correct, we give each of the 2 men 3 modia, each of the 30 women 2 modia, and each of the 68 children a 1/2 modium. This results in a total number of 100 modia:

2 men would receive 2 × 3 modia	= 6 modia
30 women would receive 30 × 2 modia	= 60 modia.
68 children would receive 68 × 1/2 modia	= 34 modia
	100 modia

The point to be emphasized here is that the above solution is based on contemporary algebraic reasoning. Even though Diophantine equations were known in Alcuin's era, the reasoning required to solve them was not necessarily a given. Like most puzzles in Alcuin's text, it is a *Gedankenexperiment,* a thought experiment that gets solvers to consider ideas in abstract terms and then follow them through to a conclusion. Incidentally, the same puzzle shows up in different guises in other parts of the world at other times, suggesting that it is also an archetype that is ensconced in real-world realities that seem to transcend geographical space and historical era. It appears much earlier in the third century CE *Bhakshali Manuscript*, discovered in northwest India in 1881. This might, of course, suggest that Alcuin may have borrowed the puzzle. But this is highly unlikely, since he would have had to know Sanskrit in an era where the language was not known broadly. Solving Diophantine puzzles requires great acumen and perseverance. Alcuin certainly knew this, realizing the difference between practical problems in calculation and puzzles as thought experiments in mathematics. Many of Alcuin's puzzles continue to find their way, in one version or other, into contemporary collections. The most famous one is the River-Crossing Puzzle (Puzzle 18), to which we shall return subsequently.

The apparent popularity of the *Propositiones* was matched by several other medieval compilations. One of these, a collection of a hundred mechanical puzzles titled *Kitab al-hiyal* ("The Book of Ingenious Devices"), was put together by the eighth-century inventor Mohammed ibn Musá ibn Shakir of Baghdad. Another, titled *The Book of Precious Things in the Art of Reckoning*, was written by the ninth-century Egyptian mathematician Abu Kamil. By the thirteenth century, such anthologies had become commonplace. Of those, Fibonacci's *Liber Abaci*, published in 1202, is undoubtedly the most well known (see Fibonacci 2002).

Fibonacci designed his book as a practical introduction to the Hindu-Arabic number system, which he had learned to use during his travels to the Middle East as a businessman. To get the system accepted in Italy, and more broadly in Europe, Fibonacci had to explain the zero concept that mystified the philosophers of his era. The 0 probably originated as far back as 600 BCE in India, although similar symbols existed in other parts of the ancient world at different times. It was the ninth-century Persian scholar, Al-Khwarizmi, who introduced the zero idea, which he called *as-sifr* "number emptiness" (a translation of the Hindu word *sunya* meaning "void" or "empty"), to Europe before Fibonacci. But his explanation went largely unnoticed until later. If "0" stood for "nothing," thinkers of the era argued, then it surely was "nothing," and thus had no conceivable uses. Fibonacci solved their dilemma by showing that 0 did indeed have a very practical function. It was no more than a convenient arithmetical sign—a "place-holder" for separating columns of figures (Posamentier and Lehmann 2007: 11).

One of the puzzles in the book has had far-reaching implications, since its solution turns up in various domains of mathematics, as well as in nature, art, and many areas of human invention. Known as the Rabbit Puzzle, it is one of the best-known puzzles of math history and is thus worthwhile presenting it briefly here. The puzzle is found in the third section of the *Liber*:

> A certain man put a pair of rabbits, male and female, in a very large cage. How many pairs of rabbits can be produced in that cage in a year if every month each pair produces a new pair which, from the 2nd month of its existence on, also is productive?

Without going into any elaborate discussion of the puzzle here, since it will be described subsequently, the answer can be displayed in terms of the following sequence of numbers, each one representing the pairs of rabbits in successive months:

1, 1, 2, 3, 5, 8, 13, 21, 34, 55, 89, 144, 233

The salient characteristic of this sequence is that each number in it is the sum of the previous two: 2 (the third number) = 1 + 1 (the sum of the previous two); 3 (the fourth number) = 1 + 2 (the sum of the previous two); and so on. Known aptly as the

Fibonacci Sequence, it can, of course, be extended ad infinitum, by applying the recursive rule of continually adding two previous numbers to generate the one after them:

1, 1, 2, 3, 5, 8, 13, 21, 34, 55, 89, 144, 233, 377, 610, 987, ...

This sequence has been found to conceal many unexpected mathematical patterns—again, validating Ahmes' observation that mathematics gives us entrance into all "obscure secrets." For example, if the nth number in the sequence is x, then every nth number after x turns out to be a multiple of x. It also shows up in Riemann's zeta function (discussed briefly above). Outside of mathematics proper, it appears in the morphology of daisies, which tend to have 21, 34, 55, or 89 petals (= the eighth, ninth, tenth, and eleventh numbers in the sequence); similarly, Fibonacci numbers describe trilliums, wild roses, bloodroots, columbines, lilies, and irises. A major chord in Western music is made up of the octave, third, and fifth tones of the scale, that is, of tones 3, 5, and 8 (another short stretch of consecutive Fibonacci numbers). The list of manifestations of Fibonacci numbers within and outside of mathematics is truly astounding. As Morris Kline (1985: 42) aptly notes, not only are we "completely ignorant about the underlying reasons" for such coincidences, but "we shall perhaps always remain ignorant of them."

Like Alcuin's *Propositiones*, Fibonacci's *Liber* was likely used as a math textbook, leading to the gradual demise of the Roman numeral system and the adoption of the decimal system as the standard. The facility with which the *Liber* clarifies complex mathematical ideas through puzzles is a corollary to the Generalization Principle above—that is, puzzles allow the mind to grasp general principles of structure that might otherwise go unnoticed. Another famous puzzle that illuminates the structure of sequences is the one by the thirteenth-century scholar Ibn Khallikan. It is paraphrased below:

> How many grains of wheat are needed on the last square of a 64-square chessboard if 1 grain is to be put on the first square of the board, 2 on the second, 4 on the third, 8 on the fourth, and so on in this fashion?

If one grain (= 2^0) of wheat is put on the first square, two grains (= 2^1) on the second, four on the third (= 2^2), eight on the fourth (= 2^3), and so on, it is obvious that 2^{63} grains will have to be placed on the sixty-fourth square. The puzzle represents, in concrete fashion, the meaning of the geometric sequence with general term 2^n. The last term, 2^{63}, can be represented as 2^{n-1}. Now, Ibn Khallikan asserted, the value of 2^{63} is so large that no chessboard could ever hold that many grains. As we shall see subsequently, this is not all that the puzzle reveals. Its solution, 2^{n-1}, crops up in other puzzles and mathematical artifacts that seem to have little in common with this one. It appears to be yet another puzzle archetype.

The puzzles of Metrodorus, Alcuin, Ibn Khallikan, Fibonacci, and other medieval and early Renaissance mathematicians were unquestionably influential in shaping mathematical history. The fact that some of their puzzles bore mathematical implications and surfaced outside of mathematics in unexpected ways reinforces the notion that puzzles may harbor archetypes serendipitously. The notion of serendipity comes initially from an ancient Persian fairy tale, *The Three Princes of Serendip*. The tale unfolds somewhat as follows. Three princes from Ceylon were journeying in a strange land when they came upon a man looking for his lost camel. The princes were perplexed, but they asked the owner: Was it missing a tooth? Was it blind in one eye? Was it lame? Was it laden with butter on one side and honey on the other? Was a pregnant woman riding it? Incredibly, the answer to all their questions was "yes." The owner accused the princes of having stolen the animal since, logically, they could not have had such precise knowledge. But the princes merely pointed out that they had observed the road on which they traveled, noticing several patterns in it: for example, the grass on either side was uneven, suggesting a lame gait; there were places where chewed food seemed to have come out of a gap in the animal's mouth; there were uneven patterns of footprints, which were signs of awkward mounting and dismounting typical of a woman who was pregnant; and there were differing accumulations of ants and flies, which congregate around butter and honey. Their questions were really prompted by inferences based on these observations. Knowing the world in which they lived, they were able to make concrete connections between the observations and what happened.

English writer Horace Walpole came across the tale and, since Serendip was Ceylon's ancient name, coined the word *serendipity*, to designate how we come about many discoveries in a similar way. The term has been applied, for example, to explain how Wilhelm Conrad Roentgen discovered X-rays by seeing their effects on photographic plates, Alexander Fleming penicillin by noticing the effects of a mold on bacterial cultures, and so on (Roberts 1989, Eco 1998, Merton and Barber 2003).

Serendipitous events result from the connective power of abduction, allowing us to figure out why something is the way it is on the basis of experience and by seeing a connectivity among things (Hofstadter 1979, Hofstadter and Sander 2013). The Italian philosopher Giambattista Vico (Bergin and Fisch 1984) was among the first to attribute serendipitous events to the imagination, or the *fantasia*, as he called it—although he did not use the term serendipity, of course, since it was coined later. As Verene (1981: 101) puts it, the Vichian *fantasia* allows humans "to know from the inside" by extending "what is made to appear from sensation beyond the unit of its appearance and to have it enter into connection with all else that is made by the mind from sensation." The debate on the imagination goes back to Plato, who separated the image (*eikon*) from the idea (*eidos*). This set in motion the tendency to view rational thought as separate from creative mental imagery, or *lógos* from *mythos*. Descartes reinforced this duality by claiming that mental images proceeded without logic, and so could not be trusted empirically. What Plato and Descartes forgot is that inventive ideas start in the imagination; they then migrate to the rational part of the brain, as argued here, to produce stable systems of knowledge.

In his classic investigation of the mathematical mind, *What is Mathematics, Really?* (1998), Reuben Hersh presents a similar type of argument. Mathematics is a human invention, Hersh argues, made by our particular kind of brain, which impels us to probe the information we take in from the world in an imaginative way. Like other thoughts, mathematical ones are actual "neural events" in the brain (Dehaene 2014) as well as historically derived mental constructs. Mathematics arises through a partnership of the biology of the brain with the forces of history. So, for instance, the neural configuration of the notion of linear equations (implicit in several of Ahmes' puzzles) comes from thinking about situations where it could be used practically to express their structural essence (Hersh 2014).

James Alexander (2012) has identified three dimensions of math knowledge that are relevant to this whole line of discussion—"pre-math," "math," and "mathematics." "Pre-math" is innate in the brain, including a primitive sense of number and space. Some animals other than humans may share the same kind of neural capacity as Dehaene (1997) and others have argued. "Math" is what we learn as a set of formal skills, from elementary school to more advanced levels of learning. It is what educators, public policy makers, mathematicians, and other authorities want every-one to be competent in. "Mathematics" is the discipline itself, with its own profes-sional culture, its research agendas and epistemologies, its sense of correctness built around proofs, and so on. The boundaries among the dimensions are fuzzy, and certainly there are many cross-influences, but the distinctions are useful nonetheless. Many early puzzles can be seen as transforming "pre-math" intuitions into "math" knowledge. The latter is often generalized concretely into "mathematics" according to the Generalization Principle (above).

Recreational Mathematics

Not all puzzles are the source of new ideas, of course. But many are. These will be the main focus of this book. The study of those puzzles that have historically generated new ideas and models of mathematical structure now falls under the rubric of *recreational mathematics*—defined as the branch that examines the theoretical implications of math puzzles.

Recreational mathematics can also be located as a branch of cognitive science, since it begs the general psychological question: What goes on in the mind of individuals as they construct or solve puzzles? As Robert Sternberg (1985) has shown with several experimental studies, solving puzzles involves the simultaneous utilization of three thinking modes. The first one, called selective encoding, refers to the process of selecting information that is relevant to the task at hand, while discarding that which is not. In so doing we "squeeze out" relevant ideas and maybe even hidden truths (see also Neuman 2007, Nave, Neuman, Howard, and Perslovsky 2014). The second, called selective comparison, entails making hunches, in order to draw a nonobvious truth between comparable pieces of information. And the third, named selective combination, implies connecting the pieces in order to form a singular thought-form that appears to describe the given situation or task at hand. Sternberg calls it a "triarchic model" of intelligence and, indeed, it seems to be an appropriate one for modeling how many puzzles are solved, being consistent with the discussion above, by and large.

To see how such thinking might unfold (hypothetically, of course), consider a well-known puzzle devised by the French Jesuit poet and scholar Claude-Gaspar

Bachet de Mézirac—a puzzle that he included in his 1612 collection titled *Problèmes plaisans et délectables qui se font par les nombres* (Bachet 1984):

> What is the least number of weights that can be used on a scale to weigh any whole number of pounds of sugar from 1 to 40 inclusive, if the weights can be placed on either of the scale pans?

We might, at first, assume that all the weights need to be used. The "brute-force" reasoning might go somewhat as follows:

- Weigh "1 pound" of sugar by putting a "1-pound weight" on the left pan, pouring sugar on the right pan until the pans balance.
- Weigh "2 pounds" of sugar by putting a "2-pound" weight on the left pan, pouring sugar on the right pan until the pans balance.
- Weigh "3 pounds" of sugar by putting a "3-pound" weight, or equivalently, the "1-pound" and the "2-pound" weights on the left pan, pouring sugar on the right pan until the pans balance.
- Continuing in this way, we could weigh any number of integral (whole-number) pounds of sugar from "1 pound" to "40 pounds."

Clearly, the task would be cumbersome physically and trivial mathematically. Bachet's puzzle is hardly that, because it requires us to use "the least number of weights," not all of them. The Aha insight comes from a hunch—we can put the weights on both pans of the scale, and so the whole task can be done ingeniously with only four weights, the "1," "3," "9," and "27" pound weights. The reason for this is remarkably simple, but nonobvious—placing a weight on the right pan, along with the sugar, is equivalent to taking its weight away from the total weight on the left pan. For example, if "2 pounds" of sugar are to be weighed, we would put a "3-pound" weight on the left pan and a "1-pound" weight on the right pan. The result is that there are "2 pounds" less on the right pan. We will get a balance when we pour the missing "2 pounds" of sugar on the right pan.

Now, this is just the start. There are various mathematical properties in the solution that are interesting in themselves, which can be unpacked through further investigation. The four weights are, upon closer scrutiny, powers of "3":

$$1 = 3^0$$
$$3 = 3^1$$
$$9 = 3^2$$
$$27 = 3^3$$

These are sufficient because the whole numbers from "1" to "40" ($=$ the required weights) turn out to be either a multiple or power of "3," or else, one more or less. Thus, each of the first forty integers can be expressed with the above four powers via addition and subtraction—mirroring the weighing system just described:

$$1 = 3^0 \qquad\qquad (= 1)$$
$$2 = 3^1 - 3^0 \qquad (= 3 - 1 = 2)$$
$$3 = 3^1 \qquad\qquad (= 3)$$

$$4 = 3^1 + 3^0 \qquad\qquad\qquad (= 3 + 1 = 4)$$
$$5 = 3^2 - 3^1 - 3^0 \qquad\qquad (= 9 - 3 - 1 = 5)$$
$$\dots$$
$$40 = 3^3 + 3^2 + 3^1 + 3^0 \qquad (= 27 + 9 + 3 + 1 = 40)$$

All we have to do is "translate" addition in the layout above as the action of putting weights on the left pan and subtraction as the action of putting weights on the right pan (along with the sugar). The puzzle thus reveals something that might escape attention about integers and the operations of addition and subtraction. It is also an experiment of triarchic thinking: (a) it involves selective encoding, since we must select information that is relevant to the puzzle, including the fact that the "least" number of weights is at play, not all of them; (b) this leads to a selective comparison, involving a comparison of the features of addition and subtraction in an imaginative way in terms of a weighing scale; and (c) through a selective combination, of the pieces of the comparison, we can now connect weights and sugar, as just surmised, in an ingenious way.

The general implications of the puzzle are explained by Petkovic (2009: 23) as follows:

> In the *Quarterly Journal of Mathematics* (1886, Volume 21) the English mathematician. P. A. MacMahon determined all conceivable sets of integer weights to weigh all loads from 1 to n. To solve this problem, he applied the method of generating functions discovered by Euler. In this way, MacMahon generalized Bachet's weight problem. Moreover, he completed the solution of the presented problem since Bachet's approach does not give all solutions. McMahon found eight solutions: $\{1_{40}\}$, $\{1, 3_{13}\}$, $\{1_4, 9_4\}$, $\{1, 3, 9_4\}$, $\{1_{13}, 27\}$, $\{1, 3_4, 27\}$, $\{1_4, 9, 27\}$, $\{1, 3, 9, 27\}$.

The gist of the following discussion is that a simple puzzle can lead to ingenious mathematical thinking, as we shall see throughout this book, and it is a thought experiment that unveils some of the features of the mathematical mind. As such, it belongs to both recreational mathematics and cognitive science.

Bachet's puzzle collection accomplished two relevant things at once for the purposes of the present discussion: it established recreational mathematics, *ipso facto*, as a branch of mathematics (although it was not named as such at the time); and it showed how puzzles stimulate the imagination, involving what Sternberg called triarchic thinking. Another widely known puzzle anthology of the era was Henry van Etten's *Mathematical Recreations Or a Collection of Sundrie Excellent Problemes out of Ancient and Modern Phylosophers Both Usefull and Recreative*, published in French in 1624 and then in English in 1633. The title may be the first time that the term "recreations" was used in reference to mathematical puzzles, foreshadowing its modern meaning. Van Etten borrowed freely from the work of his predecessors, especially from Bachet and the *Greek Anthology*, but he also introduced many innovative puzzles of his own. Shortly thereafter, in 1647, the first similarly recreational use of math puzzles in America, modeled after Bachet's and Van Etten's compilations, was published in an almanac printed by Samuel Danforth, an emigré from England (Danforth 1647).

As mentioned, the ludic brain is both a discoverer and a trickster—often blending both. The puzzle below, from the pen of Renaissance mathematician Niccolò Tartaglia, is a case-in-point:

A man dies, leaving 17 camels to be divided among his heirs, in the proportions 1/2, 1/3, 1/9. How can this be done?

Dividing up the camels in the manner decreed by the father would entail having to split up a camel. This would, of course, kill it. So, Tartaglia suggested "borrowing an extra camel," for the sake of mathematical argument, not to mention humane reasons. With 18 camels, we arrive at a practical solution: one heir was given 1/2 (of 18), or 9, another 1/3 (of 18), or 6, and the last one 1/9 (of 18), or 2. The $9 + 6 + 2$ camels given out in this way, add up to the original 17. The extra camel could then be returned to its owner. The clever Tartaglia devised his puzzle clearly as tongue-in-cheek playfulness. Nonetheless, as Petkovic (2009: 24) observes, the puzzle bears generalizable implications, as Tartaglia himself also maintained. If there are three brothers, a, b, and c, and the proportions are $1/a:1/b:1/c$, then these are solvable by the following Diophantine equation:

$$n/(n+1) = 1/a + 1/b + 1/c$$

Below is a partial set of solutions:

$$n = 7 \ (a = 2, b = 4, c = 8)$$
$$n = 11 \ (a = 2, b = 4, c = 6)$$
$$n = 11 \ (a = 2, b = 3, c = 12)$$
$$n = 17 \ (a = 2, b = 3, c = 9)$$
$$n = 19 \ (a = 2, b = 4, c = 5)$$
$$n = 23 \ (a = 2, b = 3, c = 8)$$
$$n = 41 \ (a = 2, b = 3, c = 7)$$

Many puzzles work this way. They may seem on the surface to be purely recreational or ludic, but they nonetheless bear mathematical implications. Consider Euler's *Thirty-Six Officers Puzzle* of 1779, which was both a recreational puzzle and a model for arousing significant interest in what was then a fledgling area of mathematics, namely combinatorics. The puzzle asks, simply, if it is possible to arrange 6 regiments consisting of 6 officers each of different rank in a 6×6 square so that no rank or regiment will be repeated in any row or column. Euler correctly believed that there was no solution to the puzzle, as was later proved. Euler's best-known puzzle is, however, his *Königsberg's Bridges Puzzle*, to which we shall return subsequently. The puzzle showed that it is impossible to trace a path over a network if it did not possess certain properties. Right after Euler's demonstration, mathematicians began studying paths and networks seriously. Their efforts led to the establishment of new branches of mathematics, including *topology*, which explores the properties of all kinds of networks, as well as shapes and configurations.

The first comprehensive treatise of topology, titled *Theory of Elementary Relationships*, was published in 1863. It was written by the German mathematician

Fig. 1.14 The Möbius Strip
(Wikimedia Commons)

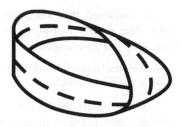

Augustus Möbius, one of the two inventors (the other being Johann Benedict
Listing) of a well-known enigmatic topological figure called the *Möbius Strip*,
which exemplifies what kinds of intriguing forms are studied within topology. The
strip has seemingly one side and one boundary—an apparent paradox, given that
strips should have two sides (a top and a bottom) and two boundaries (which delimit
the length of the strip). As is well known, it is constructed with a paper strip by
giving it a half twist, and then joining the ends to form a loop. If we draw a broken
line through the middle of one side before twisting the strip, the line runs through
that side only, not the one below. However, when twisted then the line seems to go
round and round, suggesting that there is only one surface. It is called an
unorientable object.

Whatever the explanation for this paradox, the strip has been used to study
various relations between unorientable objects and fiber bundles as well as other
topological concepts. The nineteenth century saw a further growth of interest in all
kinds of math puzzles, greatly expanding the range and scope of recreational
mathematics. In that century a famous combinatorial puzzle, known as *Kirkman's
School Girl Puzzle*—named after the notable amateur mathematician Thomas
Penyngton Kirkman, who posed it 1847—has had important implications for matrix
theory. In its rudimentary form, a matrix is a rectangular array of symbols arranged
in columns and rows in some systematic way:

> How can 15 girls walk in 5 rows of 3 each for 7 days so that no girl walks with any other girl
> in the same triplet more than once?

Solutions to this puzzle involve arranging fifteen symbols (each one representing
a specific girl) in five rows of three each within seven sets (each set corresponding to
a day of the week) so that no three specific symbols, such as numerals, appear in the
same row more than once. It is a solvable version of Euler's officers puzzle above—
suggesting that we are dealing with yet another puzzle archetype based on placement
structure that is also evident in recreational puzzles such as Sudoku.

Various solutions to Kirkman's problem have been found (Gardner 1997:
125–126). Below is one possible solution, showing how the numerals, numbered
from 0 to 14, each one representing a specific girl, can be arranged according to the
puzzle's requirements (adapted from Ball 1972: 287):

This is an example of a closed puzzle, as it has been designated in this book.
Closed puzzles may lead to more than one solution, whereas open puzzles almost
always entail a singular solution. Both are experiments in the various modalities

Monday			Tuesday			Wednesday			Thursday		
0	5	10	0	1	4	1	2	5	4	5	8
1	6	11	2	3	6	3	4	7	6	7	10
2	7	12	7	8	11	8	9	12	11	12	0
3	8	13	9	10	13	10	11	14	13	14	2
4	9	14	12	14	5	13	0	6	1	3	9

Friday			Saturday			Sunday		
4	6	12	10	12	3	2	4	10
5	7	13	11	13	4	3	5	11
8	10	1	14	1	7	6	8	14
9	11	2	0	2	8	7	9	0
14	0	3	5	6	9	12	13	1

Fig. 1.15 A solution to Kirkman's puzzle

involved in solving puzzles, including inference, analogies, comparisons, and the like. The British mathematician Augustus De Morgan, who wrote important works on the calculus and modern symbolic logic, produced a truly ingenious work in recreational mathematics, titled *A Budget of Paradoxes* (1872), in which he explored a host of mathematical theories, ideas, and suppositions that were being bandied about at the time. Along with Ahmes', Alcuin's, Fibonacci's, and Bachet's collections, among others, it can be considered a founding text in recreational mathematics.

It was the English writer and mathematician, Lewis Carroll, who raised the puzzle genre to the level of an art form. As is well known, Lewis Carroll was the *nom de plume* of Charles Lutwidge Dodgson. He created the first puzzle storybooks of history with the publication of *Pillow Problems* in 1880 (seventy-two puzzles in arithmetic, algebra, geometry, trigonometry, calculus, and probability) and *A Tangled Tale* in 1886 (puzzles originally published in monthly magazine articles). He was an eminent theorist, writing two famous treatises, *A Syllabus of Plane Algebraical Geometry* (1860) and *Euclid and His Modern Rivals* (1879), both of which were widely read by the mathematicians of his era. Carroll was fascinated by the inquisitive and fanciful imagination of children. *Alice's Adventures in Wonderland* contains all sorts of riddles and puzzles involving ingenious mind-play and double-entendre that have amused and challenged children ever since the book was first published. Carroll was captivated by the ability of puzzles to impose a peculiar kind of ordered thinking on the erratic and capricious human mind. Finding solutions to puzzles provides reassurance and a sense of order. As a teacher himself, Carroll also understood that there is no better way to stimulate the imagination than through puzzles.

By the late nineteenth century, puzzle-making was becoming a viable profession in its own right. That epoch saw, in fact, the first professional puzzlists come onto the social scene—the Americans Oswald Veblen and Sam Loyd, the Frenchman François Edouard Anatole Lucas, and the Englishman Henry E. Dudeney. Veblen is well known for his reworkings of classic puzzles, borrowing freely from such well-known collections as the *Rational Amusements for Winter Evenings* by John Jackson and the *Recreations in Mathematics and Natural Philosophy* by Edward Riddle. The latter work was an illustrated revision of a collection compiled by Jacques Ozanam (c. 1640). Veblen's work was popular because it refashioned classic puzzles for mass consumption. These could be enjoyed for their intellectual stimulation, independently of their mathematical value.

Lucas is perhaps best known as the inventor of the Tower of Hanoi Puzzle, which he published in 1883 under the pseudonym M. Claus de Siam, an anagram of his name. It is paraphrased below:

> A monastery in Hanoi has three pegs. One holds 64 gold discs in descending order of size—the largest at the bottom, the smallest at the top. The monks have orders from God to move all the discs to the third peg while keeping them in descending order. A larger disc must never sit on a smaller one. All three pegs can be used. When the monks move the last disk, the world will end. Why?

The reason the world will end is that it would take the monks 2^{64}-1 moves to accomplish the task God set before them. Even at one move per second (with no mistakes), this task will require 582,000,000,000 years. It is yet another example of a closed puzzle, hence its classification as a game as well. The puzzle will be discussed in the next chapter.

Sam Loyd was likely the first individual to earn a comfortable living from puzzle-making alone. As Matthew J. Costello (1988: 45) describes him, Loyd was "puzzledom's greatest celebrity, a combination of huckster, popularizer, genius, and fast-talking snake oil salesman." His interest in puzzles apparently started in 1860, when he became problem editor of the magazine *Chess Monthly*. By 1878, Loyd started creating chess puzzles, under the rubric of *Chess Strategy*. So popular did his puzzles become that he became convinced he could make a decent living working as a puzzlist, even though he was trained as an engineer. Working out of Manhattan, Loyd produced over ten thousand puzzles in his lifetime, most of which are extremely challenging, forcing solvers to spend countless hours trying to figure them out. His most famous and lucrative puzzle creation was the 14/15 Puzzle, which he invented in 1878. There is some dispute as to its originality, but it is clear that Loyd produced it himself, wherever he got the idea. His seemingly trivial gadget became a worldwide puzzle craze. In America, employers felt it necessary to put up notices prohibiting their employees from working on the puzzle during office hours. In France, it was characterized as a greater scourge than alcohol or tobacco. Loyd's gadget was a slim square case capable of holding 16 small square sliding blocks but containing, instead, 15 blocks, numbered from 1 to 15. The blocks were arranged in order from 1 to 13. The two numbered 14 and 15, however, were reversed. The challenge was to arrange all the blocks in order from 1 to 15 without removing any of them physically from the case: that is, by sliding one block at a time into an empty square. The insolvability of the puzzle led to various mathematical theorems and proofs—a topic that will occupy us subsequently. The roguish Loyd offered a prize of $1000 for the first correct solution, knowing full well that the puzzle could never be solved. Ironically, Loyd could not obtain a patent for his device because a solution had to be provided with the application.

Henry Ernest Dudeney was Loyd's British counterpart. At age 9, Dudeney started inventing difficult puzzles that he published in a local paper, under the appropriate pseudonym of Sphinx. In 1893, Loyd and Dudeney started a correspondence, as the first leading puzzlists of the day. But Dudeney eventually became upset with his American pen pal, breaking off relations after he started suspecting Loyd of plagiarizing his ideas. Dudeney contributed to the *Strand Magazine* for over 30 years, and he wrote a number of truly challenging puzzle books that have remained highly popular to this day.

The twentieth century witnessed a proliferation of interest in recreational mathematics. Mathematician W. W. Rouse Ball, for instance, came to be widely known for his extensive and in-depth treatment of classic puzzles, *Mathematical Recreations and Essays*, which was originally published in 1892 and reissued in at least thirteen more editions over the subsequent century. Two other widely known puzzlists, Martin Gardner and Raymond Smullyan, invented many ingenious puzzles for broad consumption, many of which nevertheless had theoretical implications within mathematics (see, for example, Gardner 1994, 1997, Smullyan 1978, 1979, 1982, 1997). Gardner was not only a prolific puzzle inventor, but also the first individual to delve into the history and meaning of puzzles—an assessment of his work can be found in Richards (1999). He wrote a puzzle column for *Scientific American* for nearly 30 years (starting in December 1956), where he popularized many of the puzzles of Loyd and Dudeney, along with those of contemporary and classic mathematicians. As Klarner (1981: iii) puts it, in the modern-day domain of puzzles, Gardner was "a force behind the scenes as well as a public figure." Smullyan was a logician who composed a host of ingenious chess and logic puzzles. He devised his first ones in 1935 when he was barely 16 years old. Smullyan's books are delectable and, in the tradition of nineteenth-century puzzlists, readable as narrative works in their own right. For example, in his brilliant book, *What Is the Name of This Book?* (1978), he takes the reader through the realm of logic and through some of its greatest conundrums, including Gödel's famous theorem, so that the reader can come to grasp the theorem in its intricacy in an entertaining manner.

A highly popular and readable book that has been influential in stressing the importance of puzzles to recreational mathematics is the one by James Kasner and John Newman, *Mathematics and the Imagination* (1940)—mentioned several times above. In it, we can see how puzzles are tied to mathematical discovery anecdotally. By solving them successfully and patiently we come away grasping intuitively that mathematics is more than a logical system of proofs; it is also part of the dialectical-ludic brain and an art in itself. Of course, not all mathematics revolves around puzzles. But it is also true that without puzzles "mathematics," in Alexander's sense, would certainly be a vastly different discipline.

In order to study puzzles in a systematic way, a basic classification is required. But this is no easy matter. The anthologies by Ahmes, Alcuin, Fibonacci, Bachet, Rouse Ball, and others present similar thematic categories—arithmetic puzzles, wieghing problems, etc. In his *Master Book of Mathematical Recreations* (1968),

Fred Schuh provides what is, perhaps, the most comprehensive classification, which includes 267 distinct puzzle types. The term recreational mathematics appeared, actually, for the first time in the 1920s in the *American Mathematical Monthly*, one of the first periodicals to provide a venue for the formal study of puzzles. As Charles Trigg (1978: 18) has aptly observed, the term "recreational" should be taken at face value—studying math through puzzles. This separates recreational mathematics from puzzles designed only for fun. Indeed, the puzzles considered to be part of recreational mathematics are hardly trivial. However, determining what puzzles are to be included under this rubric in advance is difficult. This is because, as Trigg (1978: 21) remarks: "Recreational tastes are highly individualized, so no classification of particular mathematical topics as recreational or not is likely to gain universal acceptance."

Recreational mathematics is thus a rubric under which math puzzles can be listed and examined, as well as a locus for simply classifying puzzles, even if they may not harbor any new ideas within them.

A Cognitive Flow Model

In a comprehensive treatment of the origins and evolution of puzzles, mathematical and otherwise, Helene Hovanec (1978: 10) observes that puzzles are felt to be inherently challenging because they "simultaneously conceal the answers yet cry out to be solved." It is this desire to be solved that can be used as the basis to understanding the psychology of puzzles, and thus the nature of the dialectic mind— a mind that creates questions and seeks answers, no matter how trivial the Q&A process may seem. The assessment of puzzles throughout this chapter suggests, as already mentioned, a model of mind that can be called a "cognitive flow model," whereby solving a puzzle invariably involves a cognitive flow from the imagination to abstraction and, likely, from the right hemisphere to the left hemisphere. Consider, as a case-in-point, the following puzzle, which initially presents us with an arrangement of six sticks laid out as follows, representing the fraction 1/7 in Roman numerals (I/VII):

> By moving a single, stick other than the horizontal one, change the given fraction to a fraction equal to one. See Fig. 1.16.

As with other open puzzles, the solution seems at first to be intractable. What possible Roman numeral can be constructed with these sticks that will represent the value of "one"? Only by envisioning possible arrangements of the sticks can the

Fig. 1.16 Stick puzzle

Fig. 1.17 Solution to the stick puzzle

Fig. 1.18 Dudeney's prime number magic square puzzle

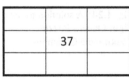

solution be found in the layout of the symbols used commonly to represent numerical values. The solution, when it comes, is yet another Aha "flash of insight." The "V" figure with one of the two upright sticks next to it in the denominator can be combined to represent a square root sign. This new arrangement will then stand for the Roman numeral "1" over "√1," and the resulting fraction is, of course, equal to one. (See Fig. 1.17).

There are several aspects of this puzzle that apply broadly to all puzzles. First, as discussed several times in this chapter, it cannot be solved in a straightforward or clearly defined way, as many problems are. Finding a solution requires envisioning what can be done with the sticks by playing a hunch. Second, as spontaneously as the solution may seem to have cropped up in our minds, after playing the hunch, it is hardly disconnected from previous experience and knowledge, as Sternberg's triarchic model implies. Only someone who is familiar with square root signs and Roman numerals, and who has reflected on the various notational practices that are used to carry out numerical representation, can envision the solution in the first place. Insight thinking and background experience are intrinsically intertwined in the solution process. The cognitive flow in this case starts with a creative envisioning of the relation between objects and numerals; after the Aha insight, the reasoning part of the mind intervenes to suggest the details of the solution.

The following classic puzzle by Henry Dudeney (2016) also exemplifies, in its own way, how the flow might occur. His puzzle is paraphrased below:

Can you arrange the following nine prime numbers {1, 7, 13, 31, 37, 43, 61, 67, 73} into an order 3 magic square? The magic constant of the square is 111. [Note that "1" is considered to be prime although this is not strictly correct.] (See Fig. 1.18).

A magic square is a square arrangement of numbers in which the sum of the rows, columns, and diagonals is constant. The answer is shown in Fig 1.19—in this case the constant is 111(as already stipulated).

Magic squares are fascinating mathematical artifacts, which will be discussed subsequently. Solving them involves a blend of imaginative thinking (envisioning arrangements in the mind) with logical reasoning (placing numbers in a systematic way). Dudeney's puzzle allows us to play with prime numbers in a challenging, yet closed, rule-based, way. So, even if it appears to "go nowhere" theoretically, it still

Fig. 1.19 Solution to
Dudeney's puzzle

67	1	43
13	37	61
31	73	7

Fig. 1.20 A solution to the
Eight Queens Puzzle
(Wikimedia Commons)

allows us to explore the dialectic mathematical mind and how it seeks to understand
structure and pattern in its own way.

The Generalization Principle above implies that the end game of recreational
mathematics is the extraction of some general principle hidden in puzzles. Consider
the so-called *Eight Queens Puzzle*, whereby eight queens must be placed on an
8-by-8 chessboard in such a way that none of the queens is able to capture any other
queen (with the normal rules of chess). The puzzle requires, in other words, that no
two queens share the same row, column, or diagonal. The puzzle statement below by
Max Bezzel in 1848 is perhaps the first one (Bennett 1910: 19):

> How does one place eight queens on an 8×8 chessboard, or, for general purposes, n queens
> on an $n \times n$ board, so that no queen is attacked by another. In addition, determine the number
> of such positions.

There are ninety-two distinct solutions to the puzzle, although if rotations and
reflections are taken into account, then it has 12 unique solutions (Bennett 1910,
Gosset 1914). Figure 1.20 is one solution.

If an array of 8 numbers, $\{k_1, k_2, \ldots k_8\}$ is used to represent the positions of the
non-attacking queens, the general solution consists of the following aspects: one
queen will be on the $k_{1\text{th}}$ square of the first column, one on the $k_{2\text{nd}}$ square of the
second column, and so on. The twelve solutions can now be represented with
corresponding numbers; for example, each of the numbers in 41582736 stand for

Fig. 1.21 Numerical
solution to the Eight Queens
Puzzle

41582736	41586372	42586137
42736815	42736851	42751863
42857136	42961357	46152837
46271358	47526138	48157263

the k_n position from the bottom in columns from left to right—the "4" stands for a queen on the fourth position from the bottom of column one; the "1" stands for a queen on the bottom position in column two; the "5" stands for a queen in the fifth position from the bottom in column three, and so on. See Fig. 1.21.

Karl Friedrich Gauss became interested in the puzzle in 1859, concluding that there were 72 solutions. Unhappy about the trial-and-error method he had used to solve the puzzle, Gauss transformed it into a problem in abstract arithmetic in order to glean any inherent general properties it may have harbored (see Foulds and Johnston 1984). To see how he accomplished this, consider the first solution above—41582736. He laid out this sequence and then put the first eight integers below it, each one representing the column in which each of the positions occurs, and adding the two:

$$
\begin{array}{cccccccc}
4 & 1 & 5 & 8 & 2 & 7 & 3 & 6 \\
\underline{1} & \underline{2} & \underline{3} & \underline{4} & \underline{5} & \underline{6} & \underline{7} & \underline{8} \\
5 & 3 & 8 & 12 & 7 & 13 & 10 & 14
\end{array}
$$

Then Gauss reversed the layout of the integers, producing the following addition:

$$
\begin{array}{cccccccc}
4 & 1 & 5 & 8 & 2 & 7 & 3 & 6 \\
\underline{8} & \underline{7} & \underline{6} & \underline{5} & \underline{4} & \underline{3} & \underline{2} & \underline{1} \\
12 & 8 & 11 & 13 & 6 & 10 & 5 & 7
\end{array}
$$

Because the eight sums are distinct integers, it can be seen that no two queens lie on the same negative diagonal (first sum) or on the same positive diagonal (second sum). The overall conclusion is that the queens will be non-attacking in the permutation 41582736. The same reasoning can be used for any permutation of queens on an $n \times n$ board, but so far no general formula has been found when n is arbitrary. In other words, this seemingly simple puzzle poses a theoretical challenge in generalization within recreational mathematics, which if solved might produce important theoretical results (Gardner 1979a). Nonetheless, the study of the Eight Queens Puzzle has had implications for graph theory, number theory, and the nature of magic squares (Watkins 2004).

The puzzle brings out the importance of the cognitive flow in mathematical thinking—a flow from a concrete situation (a chessboard) to an abstraction (the implications that it might bear). The way in which the various solutions of the puzzle are reached is not unlike the type of lateral-visualization thinking involved in solving the Nine-Dot Puzzle. It turned a concrete situation—placement of queens on a chessboard in a specific way—into abstractions about mathematical structure. As

Yair Neuman (2014: 26–27) so aptly puts it, the essence of the human mind is its ability to transform concrete impressions and experiences into abstract models and ideas:

> To identify a general pattern—a Gestalt, which is an abstraction of concrete operations—we need some kind of powerful tool that may help us to conduct the quantum leap from one level of operating in this world to another level of operating in this world. Bees, for instance, create a wonderful geometrical pattern when building their beehive. A spider weaving its web was a source of amazement for the old geometricians. Neither the bee nor the spider have ever developed the mathematical field known as Group Theory, which is the abstract formulation of "group transformations" and that can point at the deep level of similarity between different geometrical patterns.

In a fundamental sense, mathematics is the "science of abstraction." The study of how certain mathematical puzzles are related to others and of how these are "pregnant" with significance—to use Thom's term (above)—is a study in the neurology of abstraction, that is, of how we reach abstractions not in themselves but as products of an imaginative flow. This might also explain why mathematicians solve many problems and puzzles by representing them with diagrams which are concrete models that can easily be visualized and from which they can unpack otherwise hidden patterns. Diagrams permeate mathematics, both as heuristic devices in aiding problem-solving and as models or strategies used in illustrating theorems, conducting proofs, and so on. This is saliently evident in geometry where a diagram of a figure is itself an intrinsic part of a theorem or proof, guiding its logical demonstration and leading to further ideas and discoveries. But diagrams are also found in all domains of mathematics. Well-known ones include Pascal's triangle and the Sierpinski triangle. The former is a diagrammatic product of examining the layout of the terms in the binomial expansion of $(a + b)^n$ where $n = \{0, 1, 2, 3, 4, 5, \ldots\}$. The numerical coefficients in the layout can be arranged into the shape of a triangle in outline. The triangle is what probably led Pascal to make his subsequent discoveries related to the binomial expansion. The Sierpinski triangle is a model of fractal structure, with the overall shape of an equilateral triangle and subdivided recursively into smaller equilateral triangles. The triangle has been shown to be connected to Pascal's triangle and to the Tower of Hanoi puzzle (Hinz, Klavzar, Milutinovic, and Petr 2013). See Fig. 1.22.

Fig. 1.22 A Sierpinski triangle

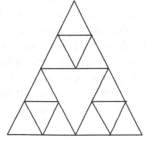

Cuts	Applied to the Circle	Pieces	Increase in Pieces
0	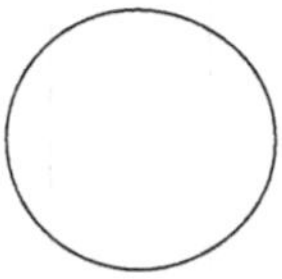	1	0
1	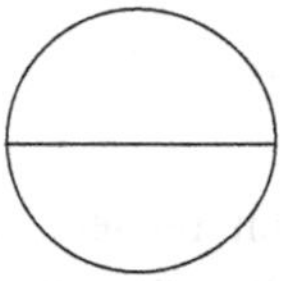	2	1
2	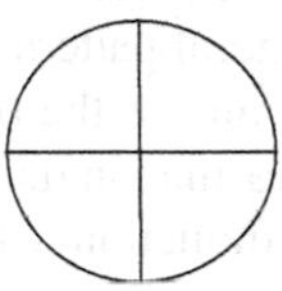	4	2
3	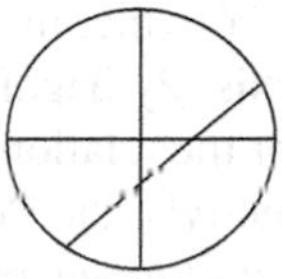	7	3

Fig. 1.23 The first three cuts in the cake-slicing puzzle

Consider as an example the following classic puzzle, which is solved diagramatically:

> What is the maximum number of pieces obtained when different cuts are made all the way across a circular cake (that is, without cutting through the same point more than twice)?

We do not need a cake, of course, to perform our cutting experiment; a circle will do, since it represents a cake in its essential outline. We start by making a number of cuts (actually, lines drawn across the circle) and record the results as we go along.

By examining Fig. 1.23 carefully a salient pattern juts out before us—the "increase in pieces" seems to parallel the increase in the "cuts." We also notice that the number of pieces increases by adding the previous number to the number of cuts. So, in the case of the 7 pieces produced by three cuts, we can see that this number equals 3 cuts +4 previous pieces. Will this pattern continue? Let's see whether it does or does not by trying out one more cut—the fourth one—and recording our information as before.

Cuts	Applied to the Circle	Pieces	Increase in Pieces
4	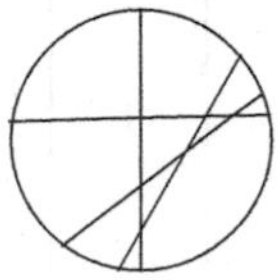	11	4

Fig. 1.24 The fourth cut in the cake-slicing puzzle

The pattern does indeed continue, as can be seen in Fig 1.24. A fourth cut produces an increase in four pieces from the previous cut, from 7 pieces to 11. Thus, a fifth cut will produce an increase in five pieces from the previous cut, that is, from 11 to 16. We can be sure that it is a general pattern via inductive proof, which implies that if the pattern applies to the nth and to the $(n\text{th} + 1)$ cases then it will occur throughout the system—like a domino that starts off a falling row of dominoes. However, the last abstraction came via induction, which initially came about as a result of concrete diagrammatic analysis.

To conclude this chapter, it is relevant to note that mathematics started out as an amalgam of numerical and geometrical analysis. Pythagoras founded a secret society that carried out the systematic investigation of the relation of numbers to geometrical diagrams—a society whose goal was to uncover the "obscure secrets" of which Ahmes spoke. This amalgam of *arithmetiké* and *geometrein* (counting and measuring) was the singular event that laid the basis for a new discipline, *mathematiké* (the science of learning). The Pythagoreans' objective was to develop an investigative intellectual tool that would allow them to explore the relation between mathematics and reality. To this amalgam the study of language can be added, as George Lakoff and Rafael Núñez (2000) so persuasively argue. As the two cognitive scientists observe, mathematicians invent their proofs and theorems through the use of analogies and metaphor, of which diagrams are visual counterparts. The idea that mathematical concepts stem from "metaphorical thinking" certainly resonates with the idea of open math puzzles as having the same structure of riddles. For Lakoff and Núñez, mathematics is an offshoot of the same neural-conceptual system that leads to the creation of language and other human skills and faculties. The fact that math puzzles arise at about the same time as riddles is anecdotal evidence of this. Whether or not this can be proven empirically, the point is that it is plausible and highly interesting and, thus, needs to be explored seriously if we are ever to come to an understanding of what mathematics is and why it leads to knowledge of the world.

Chapter 2
An Archetype Theory of Puzzles

Mathematics ought properly to be a model of logical clarity.
In actual fact there are perhaps no scientific works where you
will find more wrong expressions, and consequently wrong
thoughts, than in mathematical ones.

—Gottlob Frege (1848–1925)

As argued in the previous chapter, many classic math puzzles and games enfold some archetype as evidenced by the fact that it subsequently crops up in other domains and often becomes the basis of discovery. The archetype may show up in a "serious" exploratory puzzle or else in a "humorous" tricky one. As we have seen, there is both a serious and ludic side to the dialectic brain. In mathematics, both types of puzzles have had complementary functions as models of inherent structural features or principles. A game such as Archimedes' *loculus* is fun to play, at the same time that it is an archetype of decidability. Similarly, Tartaglia's camel puzzle is both a crafty puzzle, but also an entertaining excursion into the relation between mathematics and reality.

Dutch historian Johan Huizinga (1938) characterized the human species as *homo ludens*, in order to emphasize the role of ludic cognition in the constitution of early cultures and in the origins of consciousness. The main implication of Huizinga's theory is that we perceive life as a game and go about approaching it as such. This sentiment was captured in 1970 by the eminent British mathematician John Horton Conway with a board game that he called, appropriately, *Life* (Rucker 1987: 111–119). In the original game, Conway divided his board into squares that could be filled with a marker. Every square had eight neighboring squares, four sharing a square's sides and four diagonally adjacent, so that each marker had from 0 to 8 neighbors. The original distribution of markers was called the *first generation*. Each generation gave rise to the next according to the following three rules:

1. Every marker that has 2 or 3 neighbors stays on the board and continues to the next generation.
2. Every marker with 4 or more neighbors is removed from the board in the next generation, as are those with 0 or 1 neighbors. Such markers are said to *die out*.

© Springer International Publishing AG, part of Springer Nature 2018
M. Danesi, *Ahmes' Legacy*, Mathematics in Mind,
https://doi.org/10.1007/978-3-319-93254-5_2

3. Each empty square with exactly 3 neighbors with markers on them has a new marker placed on it in the next generation. The square is said to *come to life*.

These rules apply concurrently. Some initial configurations die out, others survive, changing and reproducing themselves. The metaphorical connection between Conway's game and human life is a transparent one. The game, like life itself, unfolds as an array of sequences of reproducing patterns. Without going into details here, it is sufficient to point out that Conway's game has had implications for algorithm design and for the theory of cellular automata. Martin Gardner (1970) made the game known widely through his *Scientific American* column, providing the following assessment:

> The game made Conway instantly famous, but it also opened up a whole new field of mathematical research, the field of cellular automata. Because of Life's analogies with the rise, fall and alterations of a society of living organisms, it belongs to a growing class of what are called "simulation games" (games that resemble real life processes).

Conway was influenced by John von Neumann's (1958) cybernetic notion of "universal constructors" or structures that could build copies or models of themselves. In Conway's game, von Neumann's models are reduced to four:

1. those that do not involve explosive growth;
2. those that possess few initial patterns with unpredictable outcomes;
3. those that have potential for emerging universal constructors;
4. those that enfold simple rules of construction.

Conway's game has also led to the notion of what computer scientist Donald Knuth (1974) calls "surreal numbers." Conway (2000) discussed the implications of such numbers within mathematical game theory, defining them as the "strengths" of positions in certain games.

Hermann Hesse's marvelous novel, *Magister Ludi* (1943), is a fictional counterpart to Conway's game. In the narrative, the meaning of life is revealed gradually to a Master of the Game, via the insights he derives from playing an archetypal bead game that involves repeating patterns, not unlike those found in Conway's game. This chapter will deal with the theory of archetypes as it pertains to puzzles. Some of the truly great puzzles of math history contain archetypes of one kind or other, and these have had substantial impacts on that history.

Alcuin's River-Crossing Puzzles

Archetype theory would explain the recurrence of such ideas in mathematics as the "squaring (polygoning) the circle" discussed in the previous chapter vis-à-vis the calculation of π or the recurrence of sequences such as the 2^n-1 or 2^{n-1} ones, which appear in several puzzles, as we have seen, including the ones by Ibn Khallikan and by Lucas (Tower of Hanoi). As Michael Schneider (1994) has cogently argued and profusely illustrated, the natural patterns and structures that recur in the universe are processed archetypally by the mind, including the presence of hexagonal structure in bee hives, the manifestation of the spiral form in many natural phenomena, the appearance of π in various natural formations, and so on. Hexagons, circles, and spirals are patterns that we discover or perceive in the world as intuitively meaningful. As a result, our mind seems impelled to encode them as archetypes, that is, as recurrent structures that have their own autonomous existence, embedding them permanently into the unconscious part where they reverberate with subsequent meaning, impelling us to search for their manifestations in new domains.

A classic example of a recurring puzzle archetype is the one found in the River-Crossing Puzzles of Alcuin, introduced briefly in the previous chapter (Hadley and Singmaster 1992, Burkholder 1993, Singmaster 1998). Support for this comes from the fact that the same puzzle appears in different guises in various languages and in various parts of the world (Ascher 1990). The situation this archetype portrays is one that might seem obvious in a practical sense, but whose abstract features might escape attention: how to devise a general back-and-forth path safely and in a minimal amount of moves. Solutions to all versions of the archetype involve critical decision-making and an assessment of possible arrangements of elements. When expressed in puzzle form, these present themselves as concrete elements of a situation that can then be generalized in some mathematical way (the Generalization Principle).

The complete title of Alcuin's collection of puzzles is *Propositiones alcuini doctoris Caroli Magni imperatoris ad acuendos juvenes*. There are four River-Crossing Puzzles in it, numbered 17, 18, 19, and 20:

Number 17: *Propositio de tribus fratribus singulas habentibus sorores*
There were three men, each having an unmarried sister, who needed to cross a river. Each man was desirous of his friend's sister. Coming to the river, they found only a small boat in which only two persons could cross at a time. How did they cross the river, so that none of the sisters were defiled by the men?
Number 18: *Propositio de homine et capra et lupo*
A certain man needed to take a wolf, a she-goat and a load of cabbage across a river. However, he could only find a boat which would carry two of these [at a time]. Thus, what rule did he employ so as to get all of them across unharmed?
Number 19: *Propositio de viro et muliere ponderantibus [plaustri pondus onusti]*
A man and his wife, each the weight of a loaded cart, who had two children each the weight of a small cart, needed to cross a river. However, the boat they came across could only carry the weight of one cart. Devise [a way] of crossing in order that the boat should not sink.
Number 20: *Propositio de hirtitiis*
A masculine and feminine [....] who had two children weighing a pound wished to cross a river.

Clearly, Puzzle 20 is incomplete and has never been reconstructed, although on the basis of the previous puzzles a reconstruction could easily be envisioned. It is to be noted that like many ancient and medieval puzzles, the wording of Puzzle 17 provides a cultural snapshot of gender relations in the medieval era—today, such a puzzle would sound antiquated or anomalous in its portrayal of these relations. Alcuin was, in all likelihood, unaware of the inbuilt gender bias in his puzzle as we would now see it; so the comment here is meant to be solely a retrospective one.

On the Original Side	On the Boat	On the Other Side
0. W̲ G̲ C̲	_ _	_ _ _
1. W̲ _ C̲	T̲ G̲ →	_ _ _
2. W̲ _ C̲	← _ T̲	_ G̲ _
3. W̲ _ _	T̲ C̲ →	_ G̲ _
4. W̲ _ _	← T̲ G̲	_ _ C̲
5. _ G̲ _	T̲ W̲ →	_ _ C̲
6. _ G̲ _	← _ T̲	W̲ _ C̲
7. _ _ _	T̲ G̲ →	W̲ _ C̲
8. _ _ _	_ _	W̲ G̲ C̲

Fig. 2.1 Alcuin's river-crossing puzzle

Puzzle 18 is the basic one—the others being essentially derivatives of the archetype it enfolds (Pressman and Singmaster 1989). The solution hinges on making the first trip over successfully. Obviously, the traveler cannot start with the cabbage, since the goat would eat the wolf; nor the wolf, since the goat would then eat the cabbage. So, his only choice is to start with the goat. Once this critical decision is made, the rest of the puzzle is solved easily. Figure 2.1 is one model of the solution path, where T = traveler, G = goat, W = wolf, and C = cabbage.

There are five general outcomes or significant patterns connected to the puzzle. First, the traveler drops off the goat (trip 1) and then at one point must go back with the goat to avoid disaster (trip 4). This requirement is not obvious in advance, but is deduced as the puzzle is solved. So, the puzzle involves both analysis and discovery in small-scale ways. Second, after the initial trip has been determined, the remaining trips can be organized slightly differently; for example, the traveler could cross over with the wolf on trip 3, rather than the cabbage, with no negative impact on the outcome. Third, the puzzle has features of both open and closed structure, which is unusual and may be the reason for its universal appeal. The puzzle is, overall, a closed one, whereby the trip structure is determined by a set of conditions and an end-state. The open aspect inheres in the initial trip; if that is not made correctly, the puzzle cannot be solved. Fourth, it is a simple model of what is now known as combinatorics, which studies the optimal ways to make arrangements of elements under given constraints. The relevant archetype can thus be called an "optimal arrangement" one (Biggs 1979). Although the study of combinations and permutations is implicit in various ancient mathematical texts, it was never really formulated as a distinct mode of inquiry. Of course, Alcuin himself does not derive the combinatorial implications from his puzzle; later mathematicians did. However, the puzzle has all the features of a combinatorial system. Some mathematical historians trace the basic ideas in combinatorics to Fibonacci's Rabbit Puzzle, but the argument made here is that the archetype underlying these ideas is to be found in Alcuin's puzzle. Fifth, since the number of trips across is required to be minimal, the puzzle also embeds a complementary archetype that can be called a "critical path" one, or the analysis of a sequence of stages determining the minimum time or moves needed for an operation.

River-crossing puzzles have appeared in a variety of cultures (Ascher 1990, Gerdes 1994). They have also surfaced in the writings of mathematicians such as

On the Original Side	On the Boat	On the Other Side
0. $\underline{H_1}\ \underline{W_1}\ \underline{H_2}\ \underline{W_2}\ \underline{H_3}\ \underline{W_3}$	$\_\ \_$	$\_\ \_\ \_\ \_\ \_\ \_$
1. $\_\ \_\ \underline{H_2}\ \underline{W_2}\ \underline{H_3}\ \underline{W_3}$	$\underline{H_1}\ \underline{W_1}\ \rightarrow$	$\_\ \_\ \_\ \_\ \_\ \_$
2. $\_\ \_\ \underline{H_2}\ \underline{W_2}\ \underline{H_3}\ \underline{W_3}$	$\leftarrow\ \_\ \underline{W_1}$	$\underline{H_1}\ \_\ \_\ \_\ \_\ \_$
3. $\_\ \_\ \underline{H_2}\ \_\ \underline{H_3}\ \underline{W_3}$	$\underline{W_1}\ \underline{W_2}\ \rightarrow$	$\underline{H_1}\ \_\ \_\ \_\ \_\ \_$
4. $\_\ \_\ \underline{H_2}\ \_\ \underline{H_3}\ \underline{W_3}$	$\leftarrow\ \_\ \underline{W_2}$	$\underline{H_1}\ \underline{W_1}\ \_\ \_\ \_\ \_$
5. $\_\ \_\ \_\ \_\ \underline{H_3}\ \underline{W_3}$	$\underline{H_2}\ \underline{W_2}\ \rightarrow$	$\underline{H_1}\ \underline{W_1}\ \_\ \_\ \_\ \_$
6. $\_\ \_\ \_\ \_\ \underline{H_3}\ \underline{W_3}$	$\leftarrow\ \_\ \underline{W_2}$	$\underline{H_1}\ \underline{W_1}\ \underline{H_2}\ \_\ \_\ \_$
7. $\_\ \_\ \_\ \_\ \underline{H_3}\ \_$	$\underline{W_2}\ \underline{W_3}\ \rightarrow$	$\underline{H_1}\ \underline{W_1}\ \underline{H_2}\ \_\ \_\ \_$
8. $\_\ \_\ \_\ \_\ \underline{H_3}\ \_$	$\leftarrow\ \_\ \underline{W_3}$	$\underline{H_1}\ \underline{W_1}\ \underline{H_2}\ \underline{W_2}\ \_\ \_$
9. $\_\ \_\ \_\ \_\ \_\ \_$	$\underline{H_3}\ \underline{W_3}\ \rightarrow$	$\underline{H_1}\ \underline{W_1}\ \underline{H_2}\ \underline{W_2}\ \_\ \_$
0. $\_\ \_\ \_\ \_\ \_\ \_$	$\_\ \_$	$\underline{H_1}\ \underline{W_1}\ \underline{H_2}\ \underline{W_2}\ \underline{H_3}\ \underline{W_3}$

Fig. 2.2 Tartaglia's river-crossing puzzle

Pacioli, Tartaglia, and others. Tartaglia became especially intrigued by number 17, inventing one of his own in which three brides and their jealous husbands had to go across the river—yet another culturally revealing wording of the puzzle. In this case the stipulation is that no wife can be left on either side or on the boat without the presence of her own husband. One possible solution is given in Fig. 2.2. There are others, of course, since the puzzle involves an end-state and thus various ways to get to it. In the solution in Fig. 2.2, H_1, H_2, H_3, W_1, W_2, and W_3 stand, respectively, for the three husbands and the three wives.

A nineteenth-century version involves three missionaries and three cannibals together. The cannibals must never be allowed to outnumber missionaries on either bank. It takes nine trips to get everyone across safely (Pressman and Singmaster 1989). As these examples show, the details of Alcuin's puzzle might vary, but the underlying archetype is always the same. As Martha Ascher (1990: 26) has aptly observed, the different cultural versions of the puzzle "are expressions of their cultures and so variations will be seen in the characters, the settings, and the way in which the logical problem is framed." An interesting variation of Russian origins is paraphrased below:

> Three soldiers have to cross a river without a bridge. Two boys with a boat agree to help the soldiers, but the boat is so small it can support only one soldier or two boys at one time. So, a soldier and a boy cannot be in the boat at the same time for fear of sinking it. Given that none of the soldiers can swim, it would seem that under these circumstances just one soldier could cross the river. Yet all three soldiers eventually end up on the other bank and return the boat to the boys. How do they do it?

It takes the following crossings:

1. Both boys go to the opposite bank and one gets off.
2. The other boy brings the boat back to the soldiers and gets off.
3. A soldier crosses the river and gets off.
4. The boy there returns with the boat.
5. When he gets to the original side, both boys get on and cross the river.
6. One boy gets off and the other returns with the boat.
7. The boy gets off and a second soldier crosses the river.

8. He gets off and the second boy returns with the boat.
9. Both boys get on and cross the river.
10. One boy gets off and the other returns with the boat.
11. He gets off and the third soldier crosses the river.
12. The soldier gets off and the second boy returns with the boat.
13. The two boys get on and cross over.

Interestingly, it is impossible to arrive at a solution under the conditions stipulated by Alcuin's puzzle 17 for four couples, unless there is an island in the middle of the river, which can be used as a temporary landing place. In other cases, the island affects neither the total number of trips nor the feasibility of a successful transfer. Sam Loyd (1959: 131–132) formulated his take on the four couples version as follows, calling it the Four Elopements Puzzle:

> The story of four elopements says that four men eloped with their sweethearts, but in carrying out their plan were compelled to cross a stream in a boat which would hold but two persons at a time. It appears that the young men were so extremely jealous that not one of them would permit his prospective bride to remain at any time in the company of any other man or men unless he was also present. Nor was any man to get into a boat alone, when there happened to be a girl alone on the island or shore, other than the one to whom he was engaged. This feature of the condition looks as if the girls were also jealous and feared that their fellows would run off with the wrong girl if they got a chance. Well, be that as it may, the problem is to guess the quickest way to get the whole party across the river according to the conditions imposed. Let us suppose the island to be in the middle of the stream. Now, tell how many minimum number of trips would the boat make to get the four couples safely across in accordance with the stipulations?

A solution is shown below, with the young men represented by A, B, C, D and their prospective brides by a, b, c, d:

	On the Original Side	On the Island	On the Other Side
Fig. 2.3 Loyd's elopement puzzle	0. A B C D a b c d	--	--
	1. A B C D c d	--	a b
	2. A B C D b c d	--	a
	3. A B C D c d	b	a
	4. C D c d	b	A B a
	5. B C D c d	b	A a
	6. B C D	b c d	A a
	7. B C D d	b c	A a
	8. D d	b c	A B C a
	9. D d	a b c	A B C
	10. D d	b	A B C a c
	11. B D d	b	A C a c
	12. d	b	A B C D a c
	13. d	b c	A B C D a
	14. d	--	A B C D a b c
	15. c d	--	A B C D a b
	0. --	--	A B C D a b c d

The fact that the four-couple version of the river-crossing archetype produces a new and unexpected structural pattern—the need to insert an island into the system—is in line with the principle of emergent structure that is now part of biology, chemistry, artificial intelligence, and other fields that focus on the nature of specific systems. It describes structures that are unknown or unplanned prior to the increase in the complexity of a given puzzle, but that emerge as the puzzle is made more complex. Emergent structures allow us to gain a better understanding of the true nature of archetypes and how they undergird recurrent forms and systems.

Ibn Khallikan's Chess Puzzle

A formula that seems to crop up time and time again in different puzzles is 2^{n-1}, suggesting that it might harbor a mathematical archetype of an unknown meaning. Consider Ibn Khallikan's famous chess puzzle discussed briefly in the previous chapter. As we saw, the solution is a geometric series, $\{2^0, 2^1, 2^2, 2^3, 2^4, \ldots, 2^{63}\}$, with each term standing for the number of grains on each successive square of the chessboard:

On the first square:	1 grain $= 2^0$ grains
On the second square:	2 grains $= 2^1$ grains
On the third square:	4 grains $= 2^2$ grains
On the fourth square:	8 grains $= 2^3$ grains
On the fifth square:	16 grains $= 2^4$ grains
On the sixth square:	32 grains $= 2^5$ grains
...	
On the sixty-fourth square:	2^{63} grains
General structure:	2^{n-1}

The last (sixty-fourth) term is "2^{n-1}." The value of "2^{64-1}" (or 2^{63}) is so large that it boggles the mind to think of what kind of chessboard could hold so many grains, not to mention where that much wheat could be found. The sixty-fourth square would contain about 1.84×10^{19} grains. This figure amounts to around 3×10^{13} bushels, which is several times the world's current annual crop of wheat. Ibn Khallikan's puzzle conceals some truly intriguing patterns, in line with the principle of emergence (above). For example, if a second chessboard is placed next to the first, then the pile on the last square (= 128th square) of the second board contains "2^{127}" grains. If we subtract the number "1" from this, "$(2^{127}-1)$" we get the following result, which, incredibly, is a prime number:

170,141,183,460,231,731,687,303,715,884,105,727

As it turns out, the formula 2^n-1 can be used to test the primality of each chessboard square. For example, the fourth square has 2^3 or 8 grains on it. If we take one away from it, 2^3-1, we get 7, which is a prime number. As is well known, primes derived in this way are called *Mersenne primes*, after the French

Fig. 2.4 Prime number
squares on Khallikan's
chessboard

2^0	2^1	2^2	2^3	2^4	2^5	2^6	2^7
2^8	2^9	2^{10}	2^{11}	2^{12}	2^{13}	2^{14}	2^{15}
2^{16}	2^{17}	2^{18}	2^{19}	2^{20}	2^{21}	2^{22}	2^{23}
2^{24}	2^{25}	2^{26}	2^{27}	2^{28}	2^{29}	2^{30}	2^{31}
2^{32}	2^{33}	2^{34}	2^{35}	2^{36}	2^{37}	2^{38}	2^{39}
2^{40}	2^{41}	2^{42}	2^{43}	2^{44}	2^{45}	2^{46}	2^{47}
2^{48}	2^{49}	2^{50}	2^{51}	2^{52}	2^{53}	2^{54}	2^{55}
2^{56}	2^{57}	2^{58}	2^{59}	2^{60}	2^{61}	2^{62}	2^{63}

mathematician Marin Mersenne who used the same formula 2^n-1 as a test of the primality of specific numbers. Applied to Khallikan's chessboard, the 2^n-1 test produces prime numbers on the squares shaded in Fig. 2.4.

For example, the first four prime numbers in the chessboard above have the following structure:

Square	Value of square	Mersenne value	Prime
Third	$2^2 = 4$	$(2^n - 1) = 2^2 - 1$	3
Fourth	$2^3 = 8$	$(2^n - 1) = 2^3 - 1$	7
Sixth	$2^5 = 32$	$(2^n - 1) = 2^5 - 1$	31
Eighth	$2^7 = 128$	$(2^n - 1) = 2^7 - 1$	127

The values of "n" for the nine Mersenne primes on the chessboard are primes themselves

Square	Value of n
Third	2
Fourth	3
Sixth	5
Eighth	7
Fourteenth	13
Eighteenth	17
Twentieth	19
Thirty-second	31
Sixty-second	61

The formula 2^n-1 was thought by Euclid to generate all the perfect numbers, reinforcing the possibility that we are in fact dealing with an archetype. A perfect number is one that equals the sum of its proper divisors, with the exception of the number itself. For example, the proper divisors of the number 6 are 1, 2, and 3. Now, if we add these together we get the number: $6 = 1 + 2 + 3$. In his *City of God*, St. Augustine argued that God took 6 days to create the world, resting on the seventh, because 6 as a perfect number symbolized the perfection of creation. The next four perfect numbers after 6 are 28, 496, 8128, and 33,550,336. So far, only seventeen perfect numbers have been discovered. The last one has 1373 digits and would fill a normal page if written out.

Euclid claimed that the formula $[2^{n-1}(2^n - 1)]$ would generate perfect numbers. But, as he suspected, it generates only even perfect numbers when the expression "$(2^n - 1)$" in it is a prime number—a fact proved by Leonhard Euler two millennia later. For example, if "$n = 2$," then "$(2^n - 1) = (2^2 - 1) = (4 - 1) = 3$." Since this is a prime number, we can now use Euclid's formula to generate a perfect number:

$$[2^{n-1}(2^n-1)] = [2^{2-1}(2^2-1)] = [2^1(4-1)] = (2)(3) = 6$$

No odd perfect numbers have ever been found. They probably do not exist.

The search for a formula for deriving all the primes, as we saw in the previous chapter, raises the question of the recurrence of an archetype of mathematical infinity as represented by sequences such as the one inherent in the formula $(2^n - 1)$. Another inherent archetype that Khallikan's puzzle evokes can be formulated as follows: What would happen if we continued with larger and larger chessboards? Will the twin formulas, 2^{n-1} and $2^n - 1$, crop up in other formats? Such coincidences raise the same type of question that came to the mind of scientist Wolfgang Pauli, when he was contemplating the discovery that the strength of the electromagnetic force between two electrons yields a constant value of 1/137. He considered this finding so amazing that he has been quoted as saying that upon arrival to heaven he would ask God one question only: "Why 1/137?" In line with Pauli's question, one could add: "Why 2^{n-1} and $2^n - 1$?"

The Tower of Hanoi Puzzle

One of the most famous puzzles of the nineteenth century is the Tower of Hanoi by Édouard Lucas (1882: 127–128), which was discussed briefly in the previous chapter. Lucas' puzzle, as we saw, has 64 discs stacked in order of size on the left-most peg, the smallest at the top and largest at the bottom, thus forming a conical shape. The object (end-state) of the game is to move the entire stack to the last (third) peg to the right, according the following simple rules: (1) only one disc may be moved at a time; (2) a disc may be moved forward or backward at any time; (3) no disc may be placed on top of a smaller disc during any move and thus on any peg. As it turns out, it would take $2^{64} - 1$ moves to accomplish the task of moving the discs as stipulated by Lucas, from the left-most peg to the right-most one. Below is a three-disc version:

Fig. 2.5 Three-disc version of the Tower of Hanoi Puzzle

Table 2.1 Moves in the Tower of Hanoi game	Discs	Number of moves required: $2^n - 1$
	1	$2^n - 1 = 2^1 - 1 = (2-1) = 1$
	2	$2^n - 1 = 2^2 - 1 = (4-1) = 3$
	3	$2^n - 1 = 2^3 - 1 = (8-1) = 7$
	4	$2^n - 1 = 2^4 - 1 = (16-1) = 15$
	5	$2^n - 1 = 2^5 - 1 = (32-1) = 31$
	6	$2^n - 1 = 2^6 - 1 = (64-1) = 63$
	7	$2^n - 1 = 2^7 - 1 = (128-1) = 127$
	...	...
	64	$2^n - 1 = 2^{64} - 1$

To keep track of the moves, it is useful to name the pegs A, B, and C and the discs in order, with 1 = smallest disc, 2 = the larger disc in the middle, and 3 = the largest disc at the bottom. The moves are as follows:

1. Move disc 1 from A to C
2. Move disc 2 from A to B
3. Move disc 1 from C to B on top of 2
4. Move disc 3 from A to C
5. Move disc 1 from B to A
6. Move disc 2 from B to C on top of 3
7. Move disc 1 from A to C on top of 2 which is itself on top of 3

It took seven moves to reach the end-state. The number of moves can be represented as $2^3 - 1$, noting that the exponent stands for the number of discs in the game. If we play the same game with four discs, we will find that the number of moves is $2^4 - 1$. It is already obvious that a general pattern is involved. If we were to play the Tower of Hanoi game with four, five, and higher numbers of discs, we would in fact find that the number of moves increases according to the general formula $2^n - 1$. Table 2.1 is a summary.

In terms of the Generalization Principle, the puzzle raises the question of determining or proving the case of n discs with $p \geq 3$ pegs. The solution is not known and the kinds of ideas that it has generated need not concern us here. Now, the surprising coincidence between this game and Ibn Khallikan's puzzle is a remarkable one indeed. The above table describes the same structure that is found in the sequence of squares in the chessboard, suggesting that both puzzles harbor the same archetype; however, it is not at all obvious what it means. We shall return to this question of puzzle archetypes in the final section of this chapter. It is sufficient to note here that both 2^{n-1} and $2^n - 1$ appear unexpectedly in puzzles and games, as well as in various areas of mathematics, as discussed vis-à-vis the perfect numbers. They seem to be descriptors of recursion as well, yet another emergent structure hidden within puzzles involving sequences.

These two recursion formulas provide not only a snapshot of the internal structure of a sequence, but may also be a key to unlocking an intrinsic feature in the brain.

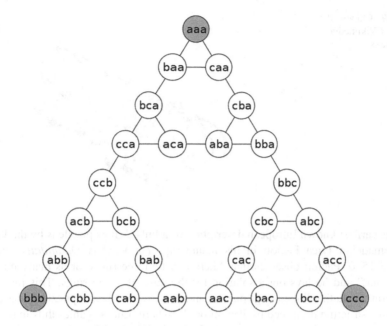

Fig. 2.6 Graph of the three-disc version (Wikimedia Commons)

However, although such snapshots may be useful for relating symbols in sequences, they hardly tell us what generates or triggers the recursive structures themselves in the first place. In computer science, recursion refers to the process of repeating items in a self-similar way and, more precisely, to a method of defining functions in which the function being defined is applied within its own definition, but in such a way that no loop or infinite chain can occur. The so-called recursion theorem says that machines can be programmed to guarantee that recursively defined functions exist. Basically, it asserts that machines can encode enough information to be able to reproduce their own programs or descriptions.

There are other fascinating patterns and implications that emerge from Lucas' puzzle, again in line with the principle of emergence. For example, by representing the game with a graph in which the nodes represent the distribution of the discs and the edges the relevant moves a connective structure emerges. For the three-disc version the graph shows a triangular structure. See Fig. 2.6.

As can be seen, this coincides with the Sierpinski triangle. Because of such structural serendipities, Lucas' game has generated a significant mathematical literature (see for example, Petkovic 2009, Hinz, Klavzar, Milutinovic, and Petr 2013). Actually, it should be noted that it has historical precedents. One of these is the Chinese Rings Puzzle, which was invented in China about 2000 years ago, but described only in the sixteenth century.

Fig. 2.7 Chinese Rings
Puzzle (Wikimedia
Commons)

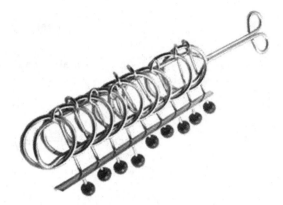

The earliest known European description of a linked rings puzzle is by the Italian mathematician Luca Pacioli in his manuscript *De Veribus Quantitatis*, written around 1510. Pacioli observes that "there can be three rings, or as many more as you want," and he lays out a solution for the case of seven rings. This raises the question of whether linked rings puzzles and the Tower of Hanoi puzzle are based on the same structure described by Pacioli or elsewhere. Lucas' games thus may be an adaptation, rather than an invention. As Petkovic (2009: 182) states:

> The solution to the Chinese rings puzzle is similar to that of the Tower of Hanoi puzzle, in that they both require a reversal of the procedure: in other words, putting the rings back on the loop. This recursion property provides an obvious link between these two puzzles. Moreover, in both cases the use of binary numbers leads to ingenious solutions, thus making these puzzles nearly identical.

The goal of the Chinese Rings Puzzle is to disentangle the long loop from the nine rings. It takes 341 moves to accomplish this end-state. Ball (1972) described a system to find the total number of moves required to remove "n" rings. The minimal number of moves is $\frac{1}{3}(2^{n+1}-1)$ if "n" is odd or $\frac{1}{3}(2^{n+1}-2)$ if "n" is even. One of the implications of the Chinese Rings Puzzle, and by extension of the Tower of Hanoi Puzzle, has been in the area of Gray-code numeration, which is an ordering of the binary number system such that two successive values differ only by one binary digit. In effect, this translates the physical movement features of the rings or discs into numerical symbols.

The Josephus Problem

Puzzles that are based on the arrangement archetype abound. They require a large dose of insight thinking to solve. Consider another well-known one called the *Josephus Problem*, after the Jewish historian Josephus of the first century CE, who supposedly saved his own life by coming up with the correct solution (Kraitchik 1942: 93–94). Here is a typical version of that puzzle:

There are fifteen Tyrants (T's) and fifteen helpless Citizens (C's) on a ship—way too many for the size of the ship. So, it is decided that the Tyrants must be thrown overboard to prevent the ship from sinking. A mythical beast, who cannot distinguish between Tyrants and Citizens, has been let loose on the ship to throw people overboard. The beast has been trained to throw over every ninth person seated in a circle. How can the Tyrants and Citizens on board be arranged in a circle so that the beast can eliminate only the Tyrants?

The beast starts at the first "C." The ninth person from the start is a "T." So he is thrown overboard. The ninth person after that is also a "T." He too is thrown overboard. The game continues in this way. This system of moves guarantees that every Tyrant is thrown overboard while all the Citizens are saved. There are various ways to arrange these in a circle, but all reveal the same kind of pattern, implying that the arrangement archetype discussed above is at play here, albeit in a different version.

Variants of the Josephus Problem are found in different cultures throughout the world—reinforcing the claim that it harbors a universal archetype. The puzzle was studied by famous mathematicians, including Leonhard Euler, because it constitutes, in puzzle form, a miniature model for investigating more complex problems in systematic arrangement. As Petkovic states (2009: 2):

> This problem, appearing for the first time in Ambrose of Milan's book ca. 370, is known as the Josephus problem, and it found its way not just into later European manuscripts, but also into Arabian and Japanese books. Depending on the time and location where the particular version of the Josephus problem was raised, the survivors and victims were sailors and smugglers, Christians and Turks, sluggards and scholars, good guys and bad guys, and so on. This puzzle attracted the attention of many outstanding scientists, including Euler, Tait, Vilf, Graham, and Knuth.

If we start going through the sequence with different numbers of people in the starting circle, we will see a few hidden patterns emerging. First, the final survivor is never someone in an even-numbered position because all of the people standing in even-numbered positions are killed first (1 kills 2, 3 kills 4, and so on). Now, any time that the starting number of people is a power of 2, the final person standing is the same as the person who started the sequence (position number 1). This is the key to figuring out where a survivor should stand. When the number of people left standing is equal to a power of 2, then it is the survivor's turn to kill a neighbor. In effect, the structure of the puzzle is recursive, suggesting that perhaps recursion itself is an archetype that manifests itself in different ways.

The Aha, Gotcha, and Eureka Effects

Solutions to the above puzzles can be seen to produce an unconscious effect, typically called the Eureka Effect, which means etymologically "I have found it," and is attributed to Archimedes, as is well known. This is because they involve a form of discovery of hidden pattern as one goes along searching to reach the end-state, rather than requiring the inferential-abductive thinking of open puzzles, which instead produces, as discussed, the Aha Effect. Some of the classic puzzles, however, produce neither effect, but rather a "Gotcha Effect," as Gardner (1982) calls it. These use various forms of trickery to bring out some hidden structural pattern or archetype, as we saw with Tartaglia's camel puzzle in the previous chapter. Gotcha puzzles are actually quite common. Consider the following one that turned up for the first time in an arithmetic textbook written by Christoff Rudolf and published in Nuremberg in 1561 (De Grazia 1981):

> A snail is at the bottom of a 30-foot well. Each day it crawls up 3 feet and slips back 2 feet. At that rate, when will the snail be able to reach the top of the well?

This puzzle requires that we connect the counting process with the actual physical situation to which it refers. Since the snail crawls up 3 feet, but slips back 2 feet, its net distance gain at the end of every day is, of course, 1 foot up from the day before. To put it another way, the snail's climbing rate is 1 foot up per day. If we conclude that the snail will get to the top of the well on the twenty-ninth day, we will have fallen into the puzzle's hidden trap. At the end of the first day, the snail will have gone up 1 foot from the bottom of the well, and will have 29 feet left to go to the top (remembering that the well is 30 feet in depth) starting on day two. The pattern up to the start of the twenty-eighth day can be summarized as follows:

Start of day	Position in the well
One	Bottom
Two	1 foot from bottom
Three	2 feet from bottom
Four	3 feet from bottom
Five	4 feet from bottom
Six	5 feet from bottom
Seven	6 feet from bottom
. . . . . .	. . . . . .
Twenty-eight	27 feet from bottom

Consider the start of day twenty-eight. As can be gleaned from the chart above, the snail would find itself at 27 feet from the bottom. This means that the snail has 3 feet to go to the top on that day. It goes up the three feet, reaches the top, and goes out, precluding its slippage back down. So, it takes the snail a full 27 days, plus a part of day twenty-eight to go up. Rather than an Aha Effect, the puzzle tends to produce a Gotcha Effect, since we can be easily duped into a miscalculation of the snail's progress. The puzzle warns us to read information carefully and extract from it the required interpretation.

A similar puzzle is found in the third section of Fibonacci's *Liber Abaci*, suggesting that it might enfold an archetype, which can be described as another kind of "path or movement archetype" consisting of optimal movements in an up-down fashion, rather than a back-and-forth one, as in Alcuin's puzzle. Fibonacci states it as follows (see statement below Fig. 2.8).

Fig. 2.8 Northrop's bookworm puzzle (from 1944)

A bookworm starts at the outside of the front cover of volume I of a certain set of books and eats his way to the outside of the back cover of volume III. If each volume is one inch thick, he must travel three inches in all. True? No, false. A moment's study of the accompanying figure shows that he has only to make his way through volume II—a distance of one inch.

A lion trapped in a pit 50 feet deep tries to climb out of it. Each day he climbs up 1/7 of a foot: but each night slips back 1/9 of a foot. How many days will it take the lion to reach the top of the pit?

The same type of Gotcha Effect is evoked by the following puzzle, due to Eugene Northrop (1944). Figure 2.8 is Northrop's presentation.

The solution hinges on developing the correct spatial perspective of the layout. We are told that the bookworm started at the outside of the front cover of Vol. I. Viewed on the page, the cover is located to the right of Vol. I. This means that the bookworm starts by burrowing through the back cover of Vol. II. When it reaches the outside of the back cover of Vol. III, it stops. That cover is to the left on the page as we view it. So, it has simply burrowed through Vol. II. The distance is, thus, 1 in. as Northrop asserts.

The Gotcha Effect is the ludic counterpart to the Aha and Eureka Effects—a psychological triad that characterizes recreational mathematics generally (see also Gardner 1979b). We shall return to the distinction among the three effects in the final chapter. For the present purpose the Eureka Effect can be defined as the effect produced by solutions that come about not instinctively or spontaneously (as in the Aha Effect) but as a result of reaching a solution through systematic thinking. Psychologists make little or no differentiation between the Aha and Eureka Effects, but the distinction is a useful one. In a manner of speaking, the Aha Effect results from "thinking outside the box"; the Eureka Effect results instead from thinking "within the box" and finding the solution ingeniously there. They characterize open and closed puzzles respectively.

The following classic puzzle devised by Lewis Carroll in 1885 is another example of Eureka thinking (Hudson 1954: 756):

There are two containers on a table, A and B. A is half full with wine, while B, which is twice A's size, is one-quarter full with wine. Both containers are filled with water and the contents are poured into a third container, C. What portion of container C's mixture is wine?

Fig. 2.9 Initial model of
Carroll's puzzle

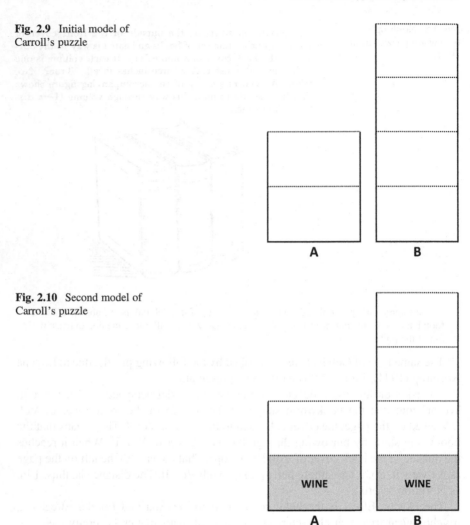

Fig. 2.10 Second model of
Carroll's puzzle

   The puzzle tells us that container A is half full with wine and that container B, which is twice the size of A, is one-quarter full with wine. First, we draw an outline model of the two containers, making B twice the size of A as shown in Fig. 2.9.

   Now, when we fill half of A and one-quarter B with wine, the containers will appear as shown in Fig. 2.10.

   Notice that there is, in actual fact, the same amount of wine in the two containers. We then fill the containers with water, as shown in Fig. 2.11.

   As can now be seen, A has two equal portions of wine and water, while B has three parts water and one part wine. Between the two containers, there are six equal parts in total—2 parts wine and 4 parts water. Logically, a mixture of these two containers will contain 2 parts wine and 4 parts water. That is, in fact, what container C will have. See Fig. 2.12.

Fig. 2.11 Third model of Carroll's puzzle

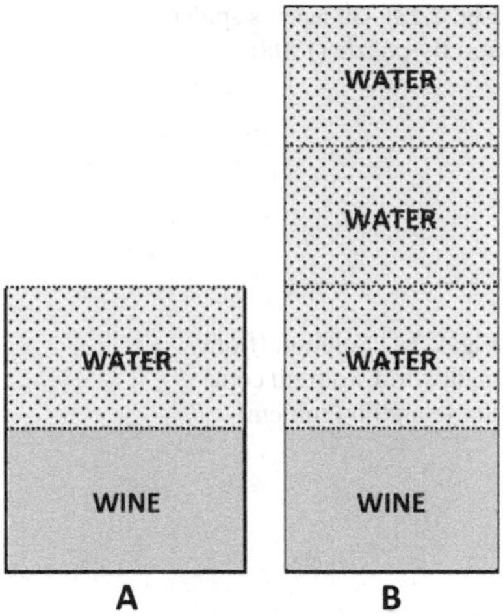

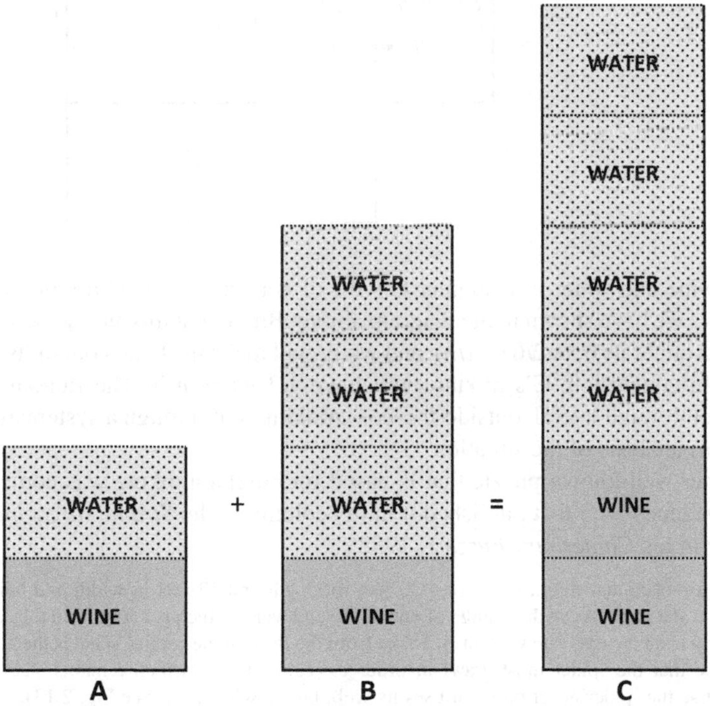

Fig. 2.12 Final model of Carroll's puzzle

Fig. 2.13 Dudeney's spider
and fly puzzle (1908)

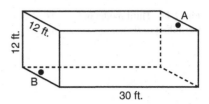

Fig. 2.14 Solution (from
mathworld.wolfram.com/
SpiderandFlyProblem)

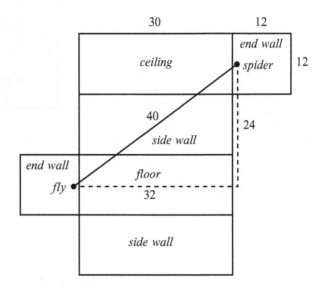

The wine and water in container C will, of course, be mixed up, not separated
neatly like we have shown in the diagram above. But in that mixture, wine will make
up 2 parts out of its 6, or 2/6 (=1/3); and water will make up 4 parts out of its 6, or 4/6
(=2/3). In conclusion, C's mixture will have 1/3 wine in it. The Eureka solution
comes not through lateral, outside-the-box thinking, but through a systematic (albeit
insightful) analysis of the situation.

Another well-known puzzle that is based on Eureka thinking was composed by
Henry Dudeney who first published it in his column in the *Strand Magazine* (1908)
and later in his *Canterbury Puzzles* (1958):

> Inside a rectangular room, measuring 30 feet in length and 12 feet in width and height, a
> spider is at a point, A, on the middle of one of the end walls, which is 1 foot from the ceiling;
> and a fly is on the opposite wall, at B, 1 foot from the floor in the centre. What is the shortest
> distance that the spider must crawl in order to reach the fly, which remains stationary?
> Of course the spider never drops or uses its web, but crawls fairly. (See Fig. 2.13).

The Eureka insight here is to "flatten the room" into its blueprint version, as
shown in Fig. 2.14:

Pursuant to this insight we can now solve this as a simple "problem" in geometry. The room has dimensions $30 \times 12 \times 12$. The spider is in one of the 12×12 walls, one foot from the ceiling. The fly is in the opposite wall, one foot from the floor. The fly remains stationary, so the spider must crawl along the walls, the ceiling, and the floor in order to capture the fly on the hypotenuse of the right triangle shown above, marking the lengths as shown. The answer is $\sqrt{(24^2 + 32^2)} = 40$. Like Carroll's puzzle, the insight comes from transforming the puzzle into a manageable diagram, rather than inferring the solution outside of the puzzle's given information.

Dudeney and Loyd were masters at producing Eureka Effects. They invented puzzles that impelled solvers to explore the structure of some part of mathematics in systematic ways. Loyd is often identified as the inventor of the *cryptarithm* (Brooke 1969), although this turns out to be incorrect, since the first documented appearance of a cryptarithm puzzle is in an 1864 issue of *The American Agriculturist*. So, the actual creator remains anonymous, even though Loyd became an ingenious creator of such puzzles. There are actually two subgenres—*cryptarithms* and *alphametics* (Hunter 1965). A cryptarithm is a puzzle in which some of the digits in an arithmetical problem have been deleted. The solver is required to reconstruct the problem by deducing numerical values on the basis of the mathematical relationships indicated by the various arrangements and locations of the numbers. Figure 2.15 is how Loyd (1959: 39) presented his puzzle.

A few numbers can be reconstructed quickly by simple inspection: (1) the 3 of the quotient, when multiplied by the 9 of the divisor, yields the product 27; so a 7 can be put in place of the asterisk under the 3; (2) this means that the last asterisk in the second row from the bottom is also 7; (3) since there is no remainder, a 4 can be inserted under the 4 in the second-to-last last row, and a 7 under the 7 in the second-to-last row; (4) the 5 of the quotient, when multiplied by the 9 of the divisor, yields the product 45; so the 5 digit can be put in place of the last asterisk of the third-to-last row from the bottom; (5) since the number subtracted from 8 is 9, the only possibility for the asterisk below it is 9.

Now, the digit above the 5 in the third-to-last row can only be 9, because there is a 4 in the row below it. This means that the second asterisk from the right in the dividend must also be replaced by a 9. Since the 3 in the quotient, when multiplied by the 9 in the divisor, yields 27, the 4 in the second-to-last row tells us that the number 3 multiplied next in the divisor—the middle asterisk—must produce a

Fig. 2.15 Loyd's cryptarithm puzzle (1959)

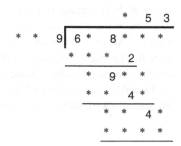

Fig. 2.16 Dudeney's
alphametic puzzle (1924)

$$
\begin{array}{cccc}
 & S & E & N & D \\
+ & & M & O & R & E \\
\hline
= & M & O & N & E & Y
\end{array}
$$

2. The only way this can be made to happen is by replacing the divisor's middle asterisk with a 4 ($3 \times 4 = 12$). This constitutes a Eureka insight, fleshed out of the puzzle via the previous systematic thinking. This means that the first asterisk in the quotient must be 8, since only 8×9 produces a number ending in 2 (72), which is the final digit in the row below the dividend. This completes the quotient. At this point the rest of the problem can be reconstructed mechanically. The answer is: $638,897 \div 749 = 853$.

Alphametics are puzzles in which numbers in an arithmetical layout are replaced by letters constituting actual words. The inventor was Dudeney. Figure 2.16 is his original alphametic (published in the July 1924 issue of *Strand Magazine*).

There is no need to go through a detailed discussion of its solution. It can instantly be established that the M at the extreme left is a carry-over digit equal to 1, because 1 is the only possibility when two digits are added together in the previous column—in this case S + M—even if that column has itself a carry-over from the column before it. That is the key Eureka insight. There are 10 digits including zero {0, 1, 2, 3, 4, 5, 6, 7, 8, 9}. The maximum that two digits can add up to is 17; this would occur when the two largest digits, 9 and 8, are added together. We note that there is no carry-over in this case because a carry-over would make the $0 = 8$, and this is impossible since that number has already been assigned hypothetically. This demonstrates that $M = 1$. The rest of the puzzle is solved with simple deduction—a mode that is based on reasoning about the structure at hand and discovering inherent principles within it. The solution is as follows: $9567 + 1085 = 10,652$.

As the above two examples show, cryptarithm and alphametic puzzles involve playing with the structure of the arithmetical operations. Solving alphametics, moreover, also entails a cross-systemic analysis, since it involves knowing the structural features of the orthographic and grammatical code in question. In French, for instance, most words end in a consonant, and therefore, as in the case of English alphametics, most numerical layouts will contain nonidentical digits in the units position. If identical digits occur, then this might suggest plural forms of nouns, adjectives, and determiners (articles, demonstratives, etc.). Italian, on the other hand, is a vocalic language. Almost all of its words end in a vowel. For this reason, identical digits appear frequently in the units position.

Perhaps the most ancient and interesting example of Eureka thinking can be found in Archimedes' Cattle Problem. As mentioned, the term *Eureka* is associated with Archimedes' discovery, while he was taking a bath, that the volume of water displaced is equal to the volume of the part of his submerged body. Legend has it that Archimedes was so thrilled by this discovery that he ran through the streets of Syracuse, naked, shouting "Eureka, Eureka!"

Archimedes dedicated his Cattle Problem to his friend, the great Alexandrian astronomer Eratosthenes. The original statement of the puzzle is lost. Of the various versions that have come down to us, the one reproduced below, taken from the authoritative English-language edition of Archimedes' works by T. L. Heath (1958: 319), contains the enigmatic extra conditions that follow the ellipsis:

> If thou art diligent and wise, O Stranger, compute the number of cattle of the Sun, who once upon a time grazed on the fields of the Thrinician isle of Sicily, divided into four herds of different colours, one milk white, another glossy black, the third yellow, and the last dappled. In each herd were bulls, mighty in number according to these proportions: understand, stranger, that the white bulls were equal to a half and a third of the black together with the whole of the yellow, while the black were equal to the fourth part of the dappled and a fifth, together with, once more, the whole of the yellow. Observe further that the remaining bulls, the dappled, were equal to a sixth part of the white and a seventh, together with all the yellow. These were the proportions of the cows: the white were precisely equal to the third part and a fourth of the whole herd of the black; while the black were equal to the fourth part once more of the dappled and with it a fifth part, when all, including the bulls, went to pasture together. Now, the dappled in four parts were equal in number to a fifth part and a sixth of the yellow herd. Finally the yellow were in number equal to a sixth part and seventh of the white herd. If thou canst accurately tell, O stranger, the number of cattle of the Sun, giving separately the number of well-fed bulls and again the number of females according to each colour, thou wouldst not be called unskilled or ignorant of numbers, but not yet shalt thou be numbered among the wise...
>
> But come, understand also all these conditions regarding the cows of the Sun. When the white bulls mingled their number with the black, they stood firm, equal in depth and breadth, and the plains of Thrinicia, stretching far in all ways, were filled with their multitude. Again, when the yellow and dappled bulls were gathered into one herd they stood in such a manner that their number, beginning from one, grew slowly greater till it completed a triangular figure, there being no bulls of other colours in their midst nor none of them lacking.
>
> If thou art able, O stranger, to find out all these things and gather them together in your mind, giving all the relations, thou shalt depart crowned with glory and knowing that thou hast been adjudged perfect in this species of wisdom.

Today, Archimedes' complicated puzzle is solved with a Diophantine method. We start our solution by letting upper-case X, Y, Z, and T stand for the number of white, black, dappled, and yellow bulls, respectively; and lower-case x, y, z, and t for the corresponding cows. The statements in the puzzle yield seven equations in eight unknowns. These equations are shown below, with fractions added and simplified:

(1) $X - T = 5/6\ Y$
(2) $Y - T = 9/20\ Z$
(3) $Z - T = 13/42\ X$
(4) $x = 7/12\ (Y + y)$
(5) $y = 9/20\ (Z + z)$
(6) $z = 11/30\ (T + t)$
(7) $t = 13/42\ (X + x)$

The answer is shown below (from Dorrie (1965: 14):

white bulls	10,366,482
black bulls	7,640,514
dappled bulls	7,358,060

yellow bulls	4,149,387
white cows	7,206,360
black cows	4,893,246
dappled cows	3,515,820
yellow cows	5,439,213

Given the magnitude of the numbers, and the complexity of the solution, it is little wonder that in classical antiquity any difficult problem was characterized as a *problema bovinum* ("cattle problem") or a *problema Archimedis* ("Archimedean problem"). The puzzle illustrates dramatically how crucial a system of representation is required to do mathematics. The solution presented here is a reductive one that precludes many structural and theoretical implications. It is used simply for the sake of illustration. The final solution was published in 1965 using a computer (Petkovic 2009: 42).

Solving Diophantine equations, in effect, brings out how Eureka thinking unfolds in general—as a way of tackling information and organizing it through symbolic compression (such as in the form of an equation) and reasoning from there. Consider Diophantus' Age Puzzle, which is actually not due to Diophantus, but comes from the *Greek Anthology* (Petrovic 2009: 10).

> Diophantus' boyhood lasted 1/6 of his life; he married after 1/7 more; his beard grew after 1/12 more, and his son was born 5 years later; the son lived to half his father's age, and the father died 4 years after the son.

By letting Diophantus' age be x, the following equation translates the statement into appropriate algebraic language:

$$x/6 + x/7 + x/12 + 5 + x/2 + 4 = x, x = 84$$

Clearly, by restating the puzzle in such language, the solution becomes tractable and easily attained. Puzzles such as these bring out the relation between representation in algebra and seemingly complex information.

Magic Squares

The lack of distinction in antiquity between numerology, or the mythical connotations of numbers, and numeration, or the use of numbers to describe the world, or between *mythos* and *lógos*, is evident in a truly remarkable ancient artifact that may constitute the world's first mathematical game or closed puzzle—the magic square, called originally Lo-Shu in Chinese. One version of the story of Lo Shu tells that there was a huge yearly flood caused by the god of the Lo River. To calm his anger, the people offered sacrifices to the god. However, the only thing that happened after each sacrifice was the appearance of a turtle from the river, which walked around nonchalantly. The people saw the turtle as a sign from the god who, they thought, kept on rejecting their sacrifices, until one time a child noticed a square on the shell

Fig. 2.17 The original
Lo-Shu

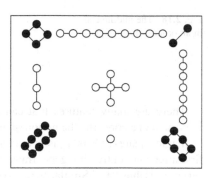

In Decimal Numbers

4	9	2
3	5	7
8	1	6

of the turtle. In it, were the first nine digits arranged in rows and columns. The child also realized that the numbers along the rows, columns, and two diagonals added up consistently to 15. From this, the people understood the number of sacrifices required of them before they could appease the god. Another version of the Lo-Shu legend has Emperor Yu the Great walking along the banks of the Lo River, when he saw a mysterious turtle crawl from the river. On its shell was a square arrangement of the first nine integers. Like the child of the previous legend, Yu noticed that the numbers in the square formed a pattern, interpreting it as a coded message from the river god.

Whichever legend is the correct one (if either), Lo-Shu was the first magic square made up of the first nine whole numbers, $\{1, 2, 3, 4, 5, 6, 7, 8, 9\}$, distributed in such a way that the three rows, three columns, and two diagonals added up to 15, known as the *magic constant*. See Fig. 2.17.

Lo-Shu spread from China to other parts of the world in the second century CE. Devising different kinds of magic squares quickly became part of occultist traditions. Like the Chinese, medieval astrologers perceived arcane divinatory properties in them, using them to cast horoscopes. The eminent astrologer Cornelius Agrippa, for example, believed that a magic square of one cell (a square containing the single digit 1) represented the eternal perfection of God. Agrippa also took the fact that a 2×2 magic square cannot be constructed to be proof of the imperfection of the four elements: air, earth, fire, and water.

The magic square is the first instance of an archetype that can be called, simply, "ordered placement." This implies that by placing objects in an arrangement in terms of structural or ordered relations hidden patterns can be detected. The same archetype shows up in other puzzles, from Kirkman's Schoolgirl puzzle (Chap. 1) to Sudoku. It is also the basis of matrix algebra and other placement systems and models.

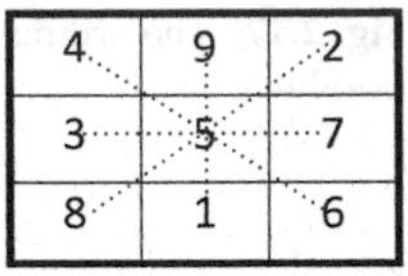

Fig. 2.18 The middle cell
in an order-3 magic square

4	9	2
3	5	7
8	1	6

There are many features that can be gleaned from magic squares, but a very important one concerns the notion of algorithm. Is there a method to the construction of magic squares? Or is it just a matter of trial and error? Lo-Shu is made from the first nine consecutive integers arranged into a square pattern. The last integer in the series, 9, is thus "3^2." Similarly, in an order "4" magic square, the last number is "4^2" ($= 16$); in an order "5" magic square, it is "5^2" ($= 25$); so in general the last integer in an order "n" square is "n^2." Using the summation technique for sequences, we can now set up an appropriate formula for the sum of the numbers in a magic square:

Sum of "n" numbers: $$\frac{n(n+1)}{2}$$

Note that Lo-Shu is an odd number square, that is, it is constructed with an odd number of integers. All such squares have a middle cell. And the number that fills that cell can be determined by figuring out in how many rows, columns, and diagonals it occurs within the square. In the case of Lo-Shu it occurs in one row, one column, and the two diagonals (four in total).

Now, there are eight possible number triplets (made up with the first nine integers) that add up to 15. These will therefore make up the rows, columns, and diagonals of the square:

$9 + 5 + 1 = 15$
$9 + 4 + 2 = 15$
$8 + 6 + 1 = 15$
$8 + 5 + 2 = 15$
$8 + 4 + 3 = 15$
$7 + 6 + 2 = 15$
$7 + 5 + 3 = 15$
$6 + 5 + 4 = 15$

We can see in Fig. 2.18 that the middle number appears in four such triplets. Eliminating the others, we are left with:

$9 + \underline{5} + 1 = 15$
$8 + \underline{5} + 2 = 15$
$7 + \underline{5} + 3 = 15$
$6 + \underline{5} + 4 = 15$

In this way we have identified the middle number—5. A similar line of reasoning can be applied to magic squares of increasing odd order. The puzzle becomes, in this way, a source for studying algorithmic structure in a system and how it can be fleshed out from arrangements. A well-known algorithm for constructing an odd order square is attributed to the mathematician Simon de la Loubère in 1693, although it is said that he probably learned about it during his travels to Asia (Andrews 1960). Let us use his algorithm on an order 5 magic square—a square consisting of the first 25 numbers, with magic constant 65. The algorithm can be laid out as follows:

1. Place 1 in the central upper cell:

		1		

2. Proceed diagonally upward to the right and place the next digit, 2, in an imaginary cell outside the actual square. Because the 2 is outside the square, bring it to the bottom of the column in alignment with it:

			2		
		1	↓		
			↓		
			↓		
			↓		
			2		

3. Put the next digit, 3, diagonally upward to the right of 2:

			2		
		1	↓		
			↓		
			↓		
			↓	3	
			2		

4. Using the same upward right diagonal movement, insert the 4 in the imaginary cell to the right of 3 and, subsequently, at the opposite end of the row:

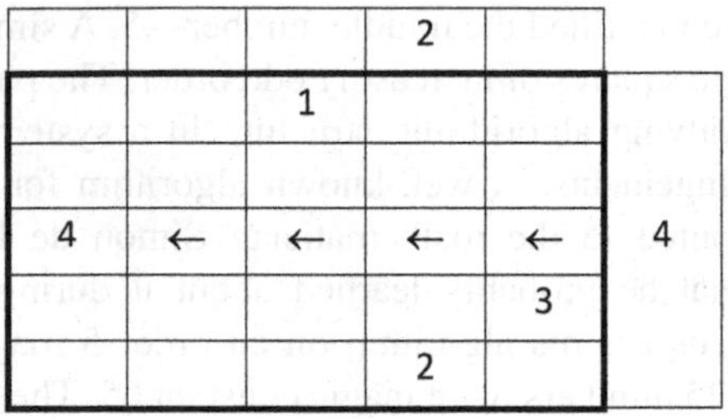

		2			
	1				
4	←	←	←	←	4
				3	
		2			

5. Insert 5 diagonally upward to the right of 4:

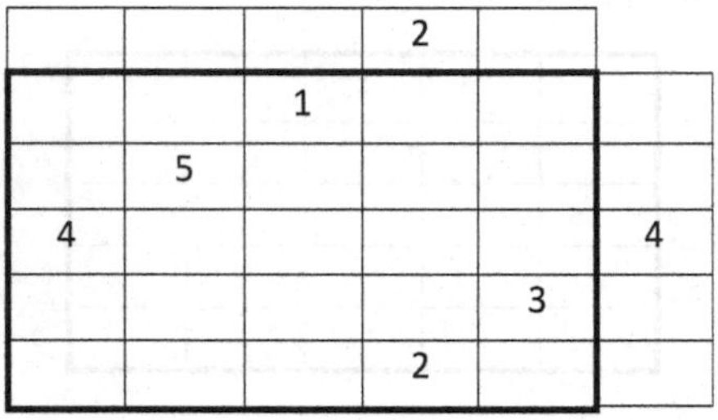

			2		
		1			
	5				
4					4
				3	
			2		

6. The same movement pattern cannot be followed to insert the 6 because the cell that is diagonally upward to the right of 5 is already occupied. The 6 is therefore written below the 5:

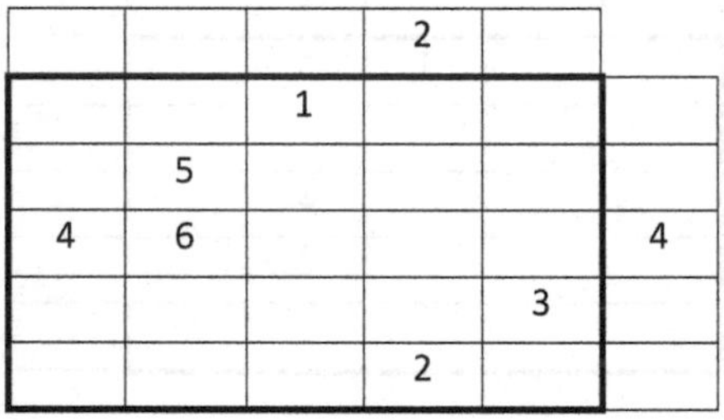

			2		
		1			
	5				
4	6				4
				3	
			2		

7. Proceed in this fashion to complete the square. See Fig. 2.19.

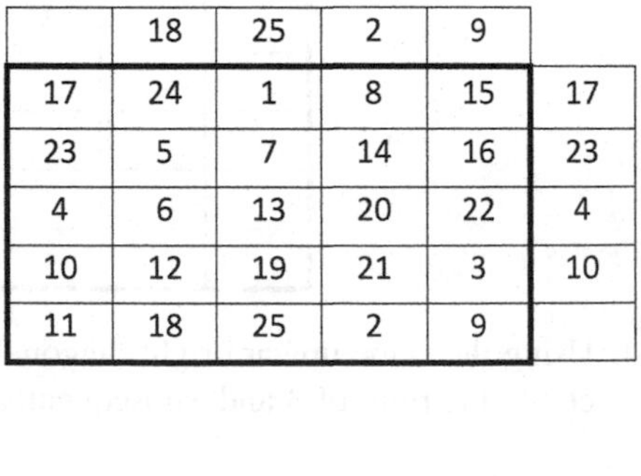

Fig. 2.19 Loubère's algorithm

	18	25	2	9	
17	24	1	8	15	17
23	5	7	14	16	23
4	6	13	20	22	4
10	12	19	21	3	10
11	18	25	2	9	

Fig. 2.20 Algorithm for an
even order square

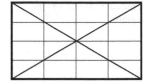

Fig. 2.21 Algorithm
continued

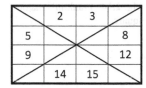

Fig. 2.22 Algorithm
completed

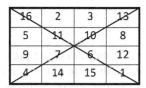

Incidentally, we can start by putting the 1 in any cell. However, this will generate a square that is magic in the rows and columns only—not in the diagonals. Now, the question becomes: Is there an algorithm for constructing even order magic squares? Figures 2.20–2.22 show a well-known one for a fourth-order square. First, we draw intersecting lines through the diagonals (Andrews 1960). See Fig. 2.20.

Next, we put in the numbers into the cells, as if they were consecutive {1, 2, 3, 4, 5, ..., 16}, leaving blank those that are crossed out by the intersecting lines. We start with 1 in the upper left corner cell. Since it is crossed, we leave it blank. We pass on to the next one to the right. Since it is empty, we put the next number in it, 2. The third cell is also empty, so we put 3 in it. The fourth cell is crossed, so we leave it empty. We proceed in this fashion until we reach the last cell in the bottom right-hand corner. See Fig. 2.21.

Now, we begin at the lower right corner, and move across the rows leftward, recording only the numbers in the cells cut by the diagonal lines. So, we start by putting 1 in the right-hand corner. The next two are filled. When we reach the bottom left corner, we put in the next number, which is 4, since 2 and 3 have already been used. We continue in this way to complete the square. See Fig. 2.22.

One of the most famous of all order-4 magic squares was constructed by the great German painter Albrecht Dürer, which he included in his famous 1514 engraving *Melancholia*. The magic square in the painting is shown below in Fig. 2.23.

Dürer's square has many hidden "magical" properties. For example, in addition to appearing in each row, column, and diagonal, the magic square constant of 34 appears as well in the locations itemized below Fig. 2.24.

Fig. 2.23 Dürer's magic
square

16	3	2	13
5	10	11	8
9	6	7	12
4	15	14	1

Fig. 2.24 Franklin's magic
square

52	61	4	13	20	29	36	45
14	3	62	51	46	35	30	19
53	60	5	12	21	28	37	44
11	6	59	54	43	38	27	22
55	58	7	10	23	26	39	42
9	8	57	56	41	40	25	24
50	63	2	15	18	31	34	47
16	1	64	49	48	33	32	17

- In the sum of the digits in the four corners ($16 + 13 + 4 + 1 = 34$)
- In the sum of the four digits in the center ($10 + 11 + 6 + 7 = 34$)
- In the sum of the digits 15 and 14 in the bottom row and the digits 3 and 2 facing them in the top row ($15 + 14 + 3 + 2 = 34$)
- In the sum of the digits 12 and 8 in the right-hand column and 9 and 5 facing them in the left-hand column ($12 + 8 + 9 + 5 = 34$)
- In the sum of the digits of each of the four squares in the corners ($16 + 3 + 5 + 10 = 34$; $2 + 13 + 11 + 8 = 34$; $9 + 6 + 4 + 15 = 34$; $7 + 12 + 14 + 1 = 34$)

One of the most extraordinary of all magic squares was the order-8 magic square devised by Benjamin Franklin. See Fig. 2.24.

Franklin's square contains a host of astonishing numerical patterns, such as the following:

- Its magic square constant is 260, and exactly half this number, 130, is the magic square constant of each of the four 4×4 squares that are quadrants of the larger square.
- The sum of any four numbers equidistant from the center is also 130.
- The sum of the numbers in the four corners plus the sum of the four center numbers is 260.
- The sum of the four numbers forming any little 2×2 square within the main square is 130. There are many more.

Not all magic squares have so many hidden patterns within them, nor do many yield themselves to algorithmic analysis (as in the cases above). For this reason, they have had implications for the larger question of decidability, an area that, as Fortnow (2013) has cogently argued, is at a key juncture of mathematics and computer science. The gist of Fortnow's argument can be summarized as follows. If we are asked to solve a 9-by-9 Sudoku puzzle, the task is a fairly simple one. The complexity of the same task arises when we are asked to solve, say, a 25-by-25 version of the puzzle. By increasing the grid to 1000-by-1000 the solution to the puzzle becomes immense in terms of effort and time.

The same can be said about magic squares, of course, or any puzzle based on a placement archetype. Computer algorithms can easily solve fairly complex Sudoku or magic square puzzles, but it becomes more and more difficult to do so as the degrees of complexity increase. The idea is, therefore, to devise algorithms that will indicate the shortest route, if any, to solving complex problems, since there would be no point in tackling a problem that may turn out not to have a solution or else is too complex to solve in a reasonable time period. If we let P stand for any problem with an easy solution, and NP for any problem with a difficult (or nonexistent) solution, then the whole question of decidability can be represented as the $P = NP$ problem. It is the most important open one in computer science and formal mathematics. It seeks to determine whether every problem whose solution can be quickly checked by computer can also be quickly solved by computer. Work on this problem has made it evident that a computer would take hundreds of years to solve some NP questions and sometimes go into a loop.

This whole line of reasoning can be traced, arguably, to Kurt Gödel's 1931 paper, which showed why in any propositional-mathematical system there is always some statement that is true, but not provable in it. Alan Turing (1936) also demonstrated a few years after Gödel's proof that if there is a general procedure to decide if a self-contained computer program will eventually halt then, paradoxically, it cannot be decided if the program will halt when it runs with that input. Turing started with the assumption that the "halting problem" was decidable and constructing a computation algorithm that halts if and only if it does not halt, which is a contradiction.

Magic squares harbor an overarching archetype, which Rucker (1987: 74) describes as the "basic notion that the world is a magical pattern of small numbers arranged in simple patterns." Euler became keenly interested in magic squares, developing his own kind of placement puzzle called a Latin Square, which has had many implications for the development of matrix algebra. It is an $n \times n$ array filled with n different symbols, each occurring exactly once in each row and exactly once in each column. Below is a 3-by-3 Latin Square in which the letters A, B, C have been inserted in each row and column so that no two identical letters occur in the same row or column. See Fig. 2.25.

Euler also constructed a "magic square of squares" in 1770. It was an order-4 square consisting of nonconsecutive square numbers in the cells with a magic square constant of 8515. See Fig. 2.26.

Fig. 2.25 Latin square

A	B	C
C	A	B
B	C	A

68^2	29^2	41^2	37^2
17^2	31^2	79^2	32^2
59^2	28^2	23^2	61^2
11^2	77^2	8^2	49^2

Examples of order-5, order-6, and order-7 magic squares of squares have since been found. Strangely, no one has yet to discover an order-3 magic square of squares, nor proved it to be impossible. Martin Gardner offered through his *Scientific American* column a one hundred dollar prize in 1996 to anyone who could devise a solution. There have been near misses, but no one has yet been able to do so. It remains an unresolved problem of recreational mathematics.

Mazes

In antiquity, buildings known as *labyrinths* with intricate interweaving rooms and corridors were imbued with mythical and mystical meanings. One of the first known labyrinths was the prison built on the island of Crete. According to legend, it was constructed by the Athenian craftsman Daedalus for King Minos to avenge the death of his son Androgeus at the hands of a group of unknown Athenians. Adding to his woe was the fact that his wife Pasiphae had fallen in love with a bull, and given birth to a half-human, half-bull beast called the Minotaur (literally, "the bull of Minos"). Embarrassed by this event, and aching to exact his revenge on Athenians generally, Minos captured and sent seven young Athenians every year into the prison. At its center he kept the voracious Minotaur, who was eager to devour anyone who ventured there. Theseus, son of King Aegeus of Athens, offered to go as one of those to be sacrificed. Ironically, Minos's daughter, Ariadne, had fallen in love with Theseus. So, to help him survive, she gave her beloved a sword with which to kill the Minotaur and a thread to mark his path through the labyrinth, so that he could find his way back out by retracing his steps on the path indicated by the string. Theseus slew the Minotaur and emerged to reunite with Ariadne, finding his way back by simply following the path marked by the thread. Aegeus had instructed Theseus to raise a white sail on his ship after he had accomplished his mission. But Theseus forgot to do so and, as legend has it, when his father saw the ship returning with black sails, he threw himself into the sea, which was thereafter called the Aegean.

No one really knows what the original Cretan Labyrinth actually looked like. Its most likely shape is found on ancient coins discovered at Knossos, the most probable site of the Cretan Labyrinth. See Fig. 2.27.

Fig. 2.27 Likely shape of
the Cretan Labyrinth

By entering at the opening, and following its single winding path, we will reach the center. This is now called a *unicursal* Eulerian graph—a graph with one path through it. Labyrinthine systems with alternative paths pose a much greater challenge, because there is no algorithm for solving them. However, some useful suggestions have been put forward by mathematicians over the years. The following are due to Lucas (1882):

1. As we go through the labyrinth, we must constantly keep looking ahead along a path to see if it ends up being a "dead end;" if so, we must avoid it and take another one at some juncture.
2. Whenever we come to a new juncture, we look ahead to scrutinize the path as open or dead.
3. If on a path we come to an old juncture or dead end, we must turn and go back the way we came.
4. We should never enter a path marked on both sides.

The Cretan Labyrinth has appealed to rulers, philosophers, mathematicians, artists, and writers alike. The later Roman emperors had copies of the labyrinth embroidered on their robes. Similar labyrinths are etched on the walls of early Christian churches.

The surrealist Argentine writer Julio Cortázar was so taken by the story that he portrayed the outside world in his novels as a phantasmal Cretan maze from which every human being must escape. Cortázar's contemporary and compatriot, Jorge Luis Borges, was also spellbound by the labyrinth concept, writing a series of truly intriguing stories, collected under the appropriate title of *Labyrinths*. The concept clearly constitutes an unconscious archetype, as defined here. It inheres in the notion that seemingly "random paths" might conceal a "hidden pattern" leading to some "secret center," which may hold the "solution" to some mystery (or obscure secret, to cite Ahmes once again). In Umberto Eco's bestselling novel, *The Name of the Rose* (1983), the central feature is a library constructed as the Cretan labyrinth. The story tells of murders committed in a medieval monastery by a serial killer. Two clerics from outside the monastery are called in to solve the murders. When first trapped within the library of the monastery, the two sleuths are able to escape with the assistance of a thread, in an obvious allusion to the Cretan myth. At the center they ultimately find the culprit—a blind monk who has poisoned the pages of Aristotle's book on humor. Fearful of the power of humor, the monk had taken it upon himself to eliminate all those who dared venture into the labyrinth's center to discover the delights and dangers of laughter.

The oldest labyrinthine design found is carved into the stone wall of a 5000-year-old grave in Sicily. Similar carvings have been discovered throughout the world. The greatest number, over three hundred, has been found in Sweden and Gotland a Swedish island in the Baltic Sea. Labyrinthine patterns have also been found on stone carvings in Ireland that go back to around 2000 BCE, in the Alps, at Pompeii, in parts of Scandinavia, Wales, England, Africa, and in Hopi rock carvings in Arizona. Many ancient buildings and cities were designed as labyrinthine structures. The Egyptian pyramids and the Christian catacombs—the networks of subterranean chambers and galleries used for burial by peoples of the ancient Mediterranean world—were designed as labyrinths, presumably to test the ability of the deceased to figure out the right path to the afterworld. The universality of the labyrinthine structure is strong indirect evidence that we are dealing with an inherent archetype of the mind (Matthews 1970)—an archetype based on the many "twists and turns" of human thought.

Complex maze puzzles started appearing after Lewis Carroll composed several ingenious ones himself, such as the one below (1958b). See Fig. 2.28.

This puzzle is clearly baffling. It seems to be chaotic, even though it conceals a hidden path. As such, it is a closed puzzle, which has a Eureka, rather than Aha, solution. Maze puzzles such as this one are clearly connected to the same kind of path archetypes found in puzzles such as the River-Crossing ones. The difference is that they are not as "neat" in graphical terms as are the graphs produced by Alcuin's puzzles and others. The idea behind mazes is to identify the optimal path in a network. Ultimately, it can be claimed that most, if not all geometric figures are graphs, and can be analyzed as such.

Fig. 2.28 Carroll's maze puzzle

Archetype Theory

Puzzle archetypes, such as placement, arrangement, optimal paths, 2^{n-1}, and others have been enlisted to explain why some puzzles lead to discovery or to uncovering emergent structure. Consider the following closed puzzle, which is found in many guises across the world and throughout the ages. The ancient Japanese called it *Hiroimono* ("things picked up"), because it is played by picking up and moving things one at a time (Costello 1988: 9).

> There are six coins in a row on a table, three white on the left and three black on the right, with one space between the two sets. Can the arrangement of the coins be reversed by moving only one coin at a time? A coin may be moved into an adjacent empty space, or jumped over one adjacent coin in an empty space. Coins may not be moved backward: that is, white coins can move only to the right and black coins to the left.

The solution clearly hinges on alternating "jumping" and "sliding" moves. Using this insight, the end-state is reached with a sequence of alternating jumping and sliding moves, summarized below (W = white coin, B = black coin). See Fig. 2.29.

This is clearly a coin version of Alcuin's River-Crossing Puzzle 17. Although anecdotal, it is fairly convincing evidence that Alcuin's archetype is embedded in many puzzle and game inventions. As is well known, the psychological concept of archetype comes from the work of Carl Jung (1983). Jung accepted Freud's basic idea of an unconscious part to the mind, but he divided it into two regions: a *personal unconscious*, containing the feelings and thoughts developed by individuals that are directive of their particular life schemes, and a *collective unconscious*, containing the feelings and thoughts developed cumulatively by the species. Jung described the latter as a "receptacle" of primordial images shared by all humanity that are beyond reflection. So, they gain expression in the symbols and forms that constitute the myths, tales, tunes, rituals, and the like that are found in cultures across the world.

Fig. 2.29 Solution to the coin puzzle

0. $W_1 W_2 W_3$ ____ $B_1 B_2 B_3$
1. $W_1 W_2$ ____ $W_3 B_1 B_2 B_3$
2. $W_1 W_2 B_1 W_3$ ____ $B_2 B_3$
3. $W_1 W_2 B_1 W_3 B_2$ ____ B_3
4. $W_1 W_2 B_1$ ____ $B_2 W_3 B_3$
5. W_1 ____ $B_1 W_2 B_2 W_3 B_3$
6. ____ $W_1 B_1 W_2 B_2 W_3 B_3$
7. $B_1 W_1$ ____ $W_2 B_2 W_3 B_3$
8. $B_1 W_1 B_2 W_2$ ____ $W_3 B_3$
9. $B_1 W_1 B_2 W_2 B_3 W_3$ ____
10. $B_1 W_1 B_2 W_2 B_3$ ____ W_3
11. $B_1 W_1 B_2$ ____ $B_3 W_2 W_3$
12. B_1 ____ $B_2 W_1 B_3 W_2 W_3$
13. $B_1 B_2$ ____ $W_1 B_3 W_2 W_3$
14. $B_1 B_2 B_3 W_1$ ____ $W_2 W_3$
15. $B_1 B_2 B_3$ ____ $W_1 W_2 W_3$

   Although he did not mention puzzles specifically, it is obvious that Jung's notion
can be used to draft an archetype theory of puzzles. Now, the question becomes:
How do puzzle archetypes spread once they have been uncovered?

   The term *meme* comes initially from Richard Dawkins' book, *The Selfish Gene*
(1976). In it, he defines memes as replicating patterns of information (ideas, laws,
clothing fashions, artworks, etc.) and of behavior that people inherit directly from
their cultural environments. Like genes, memes are passed on with no intentionality
on the part of the receiving human organism, which takes them in unreflectively
from birth, passing them on just as unreflectively to subsequent generations. The
memetic code thus parallels the *genetic code* in directing human evolution. This
clever proposal poses an obvious challenge to traditional theories of culture.
Whether it is verifiable or not, it is a useful metaphor, for the present purposes, to
explain the spread of puzzle archetypes across space and time. First, a specific puzzle
may embed a universal archetype—crossing a river or bridges, for example. The
archetype reverberates with significance, becoming a meme that is passed on from
mind to mind. From this, new mathematical ideas and even branches emerge. The
key here is that the meme is just a carrier of the archetype, which is instead the
imaginative source of the puzzle. So, the "meme theory" being adopted here is rather
different from the one put forward by Dawkins. It involves intentionality, not a lack
of it, as meme theory would generally have it.

   The discovery of combinatorics, for instance, can be traced to a specific puzzle
archetype—Alcuin's River-Crossing Puzzle. It became a puzzle meme after it was
studied by mathematicians, such as Tartaglia, who formalized its structure. In turn,
that formalization generated new branches of mathematics. So, meme theory is really
part of a general transmission theory—only those memes that are deemed to be
meaningful mathematically will be passed on and given new life and form.

Chapter 3
Puzzles and Discovery

Mathematics is as much an aspect of culture as it is a collection of algorithms.

—Carl Boyer (1906–1976)

The story behind the discovery of π has been used in this book to argue that the kind of proof devised to estimate its value harbors an archetype (polygoning the circle), since it shows up in different eras and diverse languages. Establishing a relation between archetypal thinking and puzzles has been a primary aim of this book. A second aim has been to argue that the archetype often migrates to other domains serendipitously to produce further insights both within and outside mathematics. As an example of the serendipitous appearance of π, consider the following game:

- Take a piece of cardboard and a needle.
- Mark parallel lines on the cardboard, spacing them the length of the needle apart.
- Toss the needle in the air so that it falls on the cardboard.

The object of the game is to determine the relation between the number of tosses of the needle and the number of times the needle touches a line on the cardboard. It has been found that as the number of tosses of the needle increases, the ratio of the number of tosses to the number of times the needle touches a line approaches π. This is known as Buffon's Needle Problem, first posed in the eighteenth century by Georges-Louis Leclerc, Comte de Buffon. (see Fig 3.1).

Suppose there is a floor made up of parallel lines, each one spaced apart by the same width. A needle is then dropped onto the floor. The needle is equal in length to the width space between the lines, so that it will either fall perfectly between parallel lines or else cross one of them. What is the probability that the needle will lie across a line between two strips?

Margaret Willerding (1967: 120) recounts that one of the first experiments related with the problem goes back to 1901 when "a scientist made 3408 tosses of the needle and claimed that it touched the lines 1085 times," and thus the "ratio of the number of tosses to the number of times the needle touched the lines, 3408/1085, differs from π by less than 0.1 per cent." Buffon's Needle Problem is nothing short of

© Springer International Publishing AG, part of Springer Nature 2018
M. Danesi, *Ahmes' Legacy*, Mathematics in Mind,
https://doi.org/10.1007/978-3-319-93254-5_3

Fig. 3.1 Buffon's needle problem

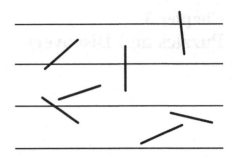

perplexing given that one can ask, legitimately: Where is the circle in this experiment? It seems that an archetype appears unexpectedly and might then lead serendipitously as an intellectual meme to discoveries and applications. In his novel *Contact* (1985), scientist Carl Sagan suggested that the creator of the universe had buried a message deep within the digits of π for us to figure out over time, recalling Ahmes' prophetic assertion that mathematics will allow us to gain entry into all "obscure secrets."

The same type of discovery story applies to the natural logarithm, e, as Ian Stewart (2008: 101) observes:

> The number e is one of those strange special numbers that appear in mathematics, and have major significance. Another such number is π. These two numbers are the tip of an iceberg—there are many others. They are also arguably the most important of the special numbers because they crop up all over the mathematical landscape.

A third aim of this book has been to argue that puzzles reveal the operation of a "dual mind," inhering in a blend of imagination and reasoning, and that the cognitive flow in puzzles is from the imagination to reasoning, not the other way around. In his "Murders in the Rue Morgue" (1841), Edgar Allan Poe called it the human being's "bi-part soul," which, he suggested, produces in all of us the mind of a "poet-mathematician." The bi-part soul is the source of all great discoveries in science and mathematics. Its operation is conspicuous in puzzles. The theme of serendipitous discovery via puzzle archetypes is the main focus of this chapter. It will examine some of the most famous puzzles that led to unexpected discoveries within and outside of mathematics, such as Fibonacci's Rabbit Puzzle and Euler's Königsberg Bridges Puzzle. It will also deal with Zeno's paradoxes and Gödel's famous theorem as instances of a clash between the imagination and reasoning or, to use Poe's term, between the two sides of the bi-part soul.

Fibonacci's Rabbit Puzzle

As discussed in Chap. 1, one of the most widely known puzzles within and outside of mathematics is Fibonacci's Rabbit Puzzle. No other puzzle has had as many serendipitous manifestations and mathematical implications as has this one (Adam 2004, Posamentier and Lehmann 2007, Devlin 2011).

There is no evidence to suggest that Fibonacci himself was aware of the significance that his puzzle would turn out to have. It was Lucas (1882) who first pointed out its emergent structural properties in the nineteenth century. Ever since, the number of properties that Fibonacci's puzzle has been found to conceal and the number of its serendipitous manifestations in nature and human life have been truly astounding. For the sake of historical accuracy, it should also be mentioned that Indian scholars were aware of the same sequence before Fibonacci (Petkovic 2009: 12). But there is no possibility that Fibonacci would have known about their work for obvious linguistic and geographical reasons. Actually, this co-occurrence of the sequence in a different era and language would bolster the idea that it harbors an archetype.

The archetype, like that in other puzzles, is recursion, in which each event in a system depends on the state attained in the previous event. The recursive pattern in the sequence can be formalized as follows:

$$F_n = F_{n-1} + F_{n-2} (F = \text{any Fibonacci number})$$

The formula provides a snapshot of the internal recursive structure of the sequence, which can be extended ad infinitum, by applying the recursive rule throughout:

$$\{1, 1, 2, 3, 5, 8, 13, 21, 34, 55, 89, 144, 233, 377, 610, 987, \ldots\}$$

Historically, it should be noted that the recursive pattern was first detected by Girard Albert in 1632. In 1753, Robert Simson discovered that, as the Fibonacci numbers increased in magnitude, the ratio between succeeding numbers approached the *golden ratio*, ϕ, whose value is 0.618..., as can be seen in the ratios of the following successive Fibonacci numbers (Dunlap 1997):

$$3/5 = 0.6$$
$$5/8 = 0.625$$
$$8/13 = 0.615$$
$$13/21 = 0.619$$
$$21/34 = 0.618$$

As we go further up the sequence, the ratios come closer and closer to the golden ratio. This is a completely unexpected emergent property of the sequence. The golden ratio itself has been found to have unexpected manifestations (Basin 1963, Livio 2002). Golden ratios appear in famous paintings, sculptures, and architectural creations. Buildings that incorporate the ratio in their design include the Parthenon (Athens, 400s BCE) and buildings designed in the 1900s by the French architect Le Corbusier. The question this elicits is the one stated throughout this book: Why would there be a connection between a sequence of numbers produced by a puzzle about copulating rabbits and one of the most enigmatic ratios in the history of human civilization? Again, this question can be contemplated in the first place if we assume that the information packed into a puzzle contains some archetype that, when unpacked, reveals emergent coincidences of structure and form in other domains.

Like the golden ratio, stretches of the Fibonacci sequence have been discovered in nature—in the spirals of sunflower heads, in pine cones, in the regular descent (genealogy) of the male bee, in the logarithmic spiral in snail shells, in the arrangement of leaf buds on a stem, in animal horns, in botanical phyllotaxis, in the arrangement of the whorls on a pinecone or pineapple, in the petals on a sunflower, in the branches of some stems, and so on and so forth. In most flowers, for example, the number of petals is: 3, 5, 8, 13, 21, 34, 55, or 89 (lilies have 3 petals, buttercups 5, delphiniums 8, marigolds 13, asters 21, daisies 34 or 55 or 89). In sunflowers, the little florets that become seeds in the head of the sunflower are arranged in two sets of spirals: one winding in a clockwise direction, the other counterclockwise. The number in the clockwise is often 21, 34 and counterclockwise 34, 55, sometimes 55 and 89.

To quote Pierre-Simon Laplace, from the introduction of his *Essai philosophique sur les probabilities* (cited in Flood and Wilson 2011: 121), the detection of such serendipities baffles the mind. Perhaps, as he implies, the human mind is designed to understand connections among things:

> We may regard the present state of the universe as the effect of its past and the cause of its future. An intellect which at a certain moment would know all forces that set nature in motion, and all positions of all items of which nature is composed, if this intellect were also vast enough to submit these data to analysis, it would embrace in a single formula the movements of the greatest bodies of the universe and those of the tiniest atom; for such an intellect nothing would be uncertain and the future just like the past would be present before its eyes.

Searching for an overarching formula to explain everything has been a dream of humanity from ancient mathematics and science to the present time (Davis and Hersh 1986, Crilly 2011: 104). Serendipity is inference by analogy, which involves figuring out why something is the way it is on the basis of experience and by seeing a resemblance among things. The power of analogy in mathematics has been discussed extensively (for example, Hofstadter 1979, Hofstadter and Sander 2013). Einstein certainly understood this to be a law of the brain when he resorted to analogies both to present his theory of relativity and to explore its profound implications. It is relevant to revisit one of these analogies here for the sake of illustration. Suppose we are on a smoothly running train moving at a constant velocity. In the train we may drop a book, or throw a ball to someone else back and forth. The book will appear to fall straight down when it is dropped; the ball will appear to travel directly from the thrower to the catcher. All these activities can be carried on in much the same way and with the same results on the ground outside the train. As long as the train runs smoothly, with constant velocity, none of these activities will be affected by its motion. On the other hand, if the train stops or speeds up abruptly, the structure of the activities will change. A book may be jarred from a seat and fall without being dropped. A ball will move differently. In effect, the laws of motion are the same for an observer in a smoothly moving train as they are for the observer on the ground. In more technical language: If two systems move uniformly *relative* to each other, then the laws of physics are the same in both systems.

Table 3.1 Fibonacci number patterns

F_n	F_{n+1}	$\rightarrow$	$F_n{}^2$	+	$F_{n+1}{}^2$	=	Fibonacci number
2	3	$\rightarrow$	$4\,(=2^2)$	+	$9\,(=3^2)$	=	13
3	5	$\rightarrow$	$9\,(=3^2)$	+	$25\,(=5^2)$	=	34
5	8	$\rightarrow$	$25\,(=5^2)$	+	$64\,(=8^2)$	=	89
8	13	$\rightarrow$	$64\,(=8^2)$	+	$169\,(=13^2)$	=	233
13	21	$\rightarrow$	$169\,(=13^2)$	+	$441\,(=21^2)$	=	610
21	34	$\rightarrow$	$441\,(=21^2)$	+	$1156\,(=34^2)$	=	1597
...	...	...	...	...	...	...	...

Fig. 3.2 Pascal's triangle and Fibonacci numbers (Wikimedia Commons)

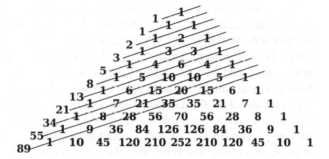

To study the serendipitous occurrences of the Fibonacci sequence as well as its hidden mathematical properties a Fibonacci Society and journal, *The Fibonacci Quarterly*, were established in 1962, by Verner Emil Hoggart. In effect, hidden properties, structures, and interrelations among the numbers in the sequence continue to be found without end. As an example, Table 3.1 above shows that the sum of the squares of two consecutive Fibonacci numbers produces another Fibonacci number—a truly remarkable hidden pattern within the sequence.

One of the most intriguing discoveries is the relation that the sequence has to Pascal's triangle, which consists of a triangular arrangement of numbers whereby a number in a given row is the sum of two numbers immediately above it in the triangle. As it turns out, the diagonal sums of the numbers in Pascal's triangle correspond to the numbers in the Fibonacci sequence {1, 1, 2, 3, 5, 8, 13, 21, 34, 55, 89, ...} (Vernadore 1991). See Fig. 3.2.

Structural serendipities such as this one have been researched extensively. It is not always clear what they mean, however. They are fascinating in themselves. As another example, consider the sum of the first ten consecutive Fibonacci numbers in the sequence:

$$1 + 1 + 2 + 3 + 5 + 8 + 13 + 21 + 34 + 55 = 143 \tag{1}$$

The sum turns out to be divisible evenly by eleven ($143/11 = 13$). Now, amazingly, the same result holds for the sum of any ten consecutive Fibonacci numbers. Take, for example, the ten that start with 55 in the sequence:

$$55 + 89 + 144 + 233 + 377 + 610 + 987 + 1,597 + 2,584 + 4,181$$
$$= 10,857 \tag{2}$$

This sum is also divisible by 11 (10,857/11 = 987). If we examine such cases more closely, it turns out that the sum of any ten consecutive numbers is equal to 11 times the seventh number in the chosen ten-digit sequence. In (1) above the seventh number is 13 and $13 \times 11 = 143$; and in (2) the seventh number is 987 and $987 \times 11 = 10,857$. The meaning of this type of coincident pattern is unknown. Returning to Ahmes, we may be confronted by an obscure secret whose meaning is mysterious.

As mathematicians started to see Fibonacci numbers appear in the most unexpected places and in surprising ways, they became interested in finding an efficient method for calculating any Fibonacci number. In principle, this is not a difficult generalization problem. To identify the 100th Fibonacci number, for instance, all we have to do is add the 98th and 99th numbers together. However, this still means we have to identify all the numbers up to the 98th, which can prove to be quite tedious. So, in the middle of the nineteenth century, the French mathematician Jacques Binet devised a formula, based on the calculations of Leonhard Euler and Abraham de Moivre beforehand, that allows us to find any Fibonacci number, if its position in the sequence is known. The Binet formula is given below:

$$F_n = \frac{1}{\sqrt{5}} \left(\left(\frac{1 + \sqrt{5}}{2} \right)^n - \left(\frac{1 - \sqrt{5}}{2} \right)^n \right)$$

It is beyond the purpose here to explain how Binet arrived at the formula. Suffice it to say that it is based on the golden ratio, formally showing the connection between Fibonacci numbers and this ratio.

If any recursive series is constructed, patterns jut out from it unexpectedly, suggesting that recursion is a pervasive archetype that may harbor within it all kinds of obscure secrets. Lucas, for instance, created his own recursive series, known as the *Lucas sequence* (Vajda 1989). He started it with the terms 2 and 1, rather than 1 and 1:

$$\{2, 1, 3, 4, 7, 11, 18, 29, 47, 76, 123, \ldots\}$$

Without going into specifics here, it is sufficient to point out that this sequence possesses a host of relationships that connect it mathematically to the Fibonacci sequence; moreover, it has had implications for congruence relations, for primality testing, among many other mathemtical constructs.

Perhaps the most significant lesson to be learned from the Fibonacci puzzle in mathematical history is that discoveries cannot be forced by logical analysis or intent (Davis and Hersh 1986). But discovery is not totally random either. It is tied to unconscious archetypal processes that humans convert into symbolic artifacts, such as puzzles. The latter then function as mental maps for further exploration, leading to subsequent discoveries.

Euler's Königsberg Bridges Puzzle

As discussed briefly in the opening chapter, Leonhard Euler created puzzles to examine mathematical ideas, such as his Thirty-Six Officers Puzzle. His most famous one is derived actually from a real situation. In the German town of Königsberg runs the Pregel River. In the river are two islands, which are connected with the mainland and with each other by seven bridges. The residents of the town would often debate whether or not it was possible to take a walk from any point in the town, cross each bridge once and only once, and return to the starting point. No one had found a way to do it but, on the other hand, no one could explain why it seemed to be impossible. Euler formulated the relevant conundrum in a famous 1736 paper that he presented to the Academy in St. Petersburg, Russia, and which he published in 1741. The puzzle it presents can be paraphrased as follows:

> In the town of Königsberg, is it possible to cross each of its seven bridges over the Pregel River, which connect two islands and the mainland, without crossing over any bridge twice?

As can be discerned, we are dealing here once again with an archetype involving paths in order to determine if the network is solvable. Euler went on to prove that it is impossible to trace a path over the bridges without crossing at least one of them twice. This can be shown by reducing the map of the area to graph form, restating the puzzle as follows:

> Is it possible to trace the following graph without lifting pencil from paper, and without tracing any edge twice? See Fig. 3.3.

This version represents the landforms as a network of vertices, and portrays the bridges as paths or edges. It shows, in its outline, that solving the puzzle is impossible without doubling back at some point. Creating more complex networks, with more and more paths and vertices in them, will reveal that it is not possible to traverse a network that has more than two odd vertices in it without having to double back over some of its paths. Euler proved this in a remarkably simple way. A network can have any number of even paths in it, because all the paths that converge

Fig. 3.3 Graph version of Euler's puzzle (Wikimedia Commons)

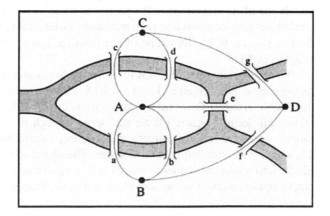

at an even vertex are "used up" without having to double back on any one of them. For example, at a vertex with just two paths, one path is used to get to the vertex and another one to leave it. Both paths are thus used up without having to double back over either one of them. In a four path network, when we get to a vertex we can exit via a second path. Then, a third path brings us back to the vertex, and a fourth one gets us out. All paths are used up.

The same reasoning applies to any even vertex network. In an odd vertex network, on the other hand, there will always be one path that is not used up. For example, at a vertex with three paths, one path is used to get to the vertex and another one to leave it. But the third path can only be used to go back to the vertex. To get out, we must double back over one of the three paths. The same reasoning applies to any odd vertex network. Therefore, a network can have, at most, two odd vertices in it. And these must be the starting and ending vertices. The relevant point here is that Euler's graph makes it possible to look at the relationships among elemental geometric systems in order to determine their structure and their implications. Euler did this by turning the puzzle into a generalizable form (Richeson 2008: 107):

> The solution to the Königsberg bridge problem illustrates a general mathematical phenomenon. When examining a problem, we may be overwhelmed by extraneous information. A good problem-solving technique strips away irrelevant information and focuses on the essence of the situation. In this case details such as the exact positions of the bridges and land masses, the width of the river, and the shape of the island were extraneous. Euler turned the problem into one that is simple to state in graph theory terms. Such is the sign of genius.

The puzzle also had implications for studying mathematical impossibility. The ancient Greeks grappled constantly with the concept of impossibility, wondering why, for example, it was seemingly impossible to trisect an angle with compass and ruler, given that bisection was such a simple procedure. For years, mathematicians attempted trisection with compass and ruler, but always to no avail. The demonstration that it was impossible had to await the development and spread of Descartes' method of converting every problem in geometry into a problem in algebra. The proof that trisection was impossible was based on the Cartesian method. It came in the nineteenth century after mathematicians had established that the equation which corresponds to trisection must be of degree 3—that is, it must be an equation in which one of its variable is to the power of 3: $ax^3 - bx^2 + cx = 0$. A construction carried out with compass and ruler translates instead into an equation to the second degree: $x^2 - a = 0$. The formal proof was published by mathematician Pierre Laurent Wantzel in 1837.

The impossibility meme crops up in many forms and guises within recreational mathematics. For instance, Loyd's 14/15 Puzzle, as mentioned in Chap. 1, is impossible to solve. To revisit his device here, Loyd put 15 consecutively numbered sliding blocks in a square plastic tray large enough to hold 16 such blocks. The blocks were arranged in numerical sequence, except for the last two, 14 and 15, which were installed in reverse order. The object of the game is to arrange the blocks into numerical sequence from 1 to 15, by sliding them, one at a time, into an empty square, without lifting any block out of the frame. See Fig. 3.4.

Fig. 3.4 Loyd's 14/15
Puzzle (Wikimedia
Commons)

Fig. 3.5 Solvable version
of the 14/15 Puzzle
(Wikimedia Commons)

As it turns out, it is impossible to accomplish this. Now, when the blocks are placed in numerical order—each one followed by a block that is exactly one digit higher—the blocks can be scrambled and the puzzle can then be easily solved. See Fig. 3.5.

Remarkably, the mathematical structure of this puzzle mirrors that of the Königsberg Bridges Puzzle, even though on the surface it does not seem to be related to it. Every instance of a block followed by one that is lower than itself is called an inversion. If the sum of all the inversions in a given arrangement is even, the solution to the puzzle is a possible one. If the sum is odd, it is impossible, just like the structire of the vertices in Euler's problem. Loyd's game had only 1 inversion (15 is followed by 14). This is an odd number, and thus it is impossible to arrange the blocks in numerical order. Loyd's game and Euler's puzzle are clearly based on the same archetype involving movement within a network—moving from one position to another in decidable ways.

In line with the Generalization Principle, it comes as little surprise that Euler himself discovered several fundamental properties about networks and graphs pursuant to his puzzle. Any geometrical figure can be re-imagined as a graphical

Fig. 3.6 The cube as a
graphical network

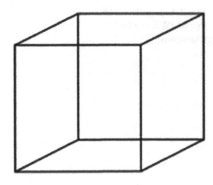

Fig. 3.7 The tetrahedron as
a graphical network

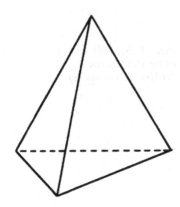

network, since it has edges, vertices, and faces. Euler found that if we subtract the
number of edges ("*e*") from the number of vertices ("*v*") and then add the number of
faces ("*f*") we will always get "2" as a result: $v - e + f = 2$.

Take, as a case-in-point, the cube figure shown in Fig. 3.6.

It has 8 vertices (sharp corners), 12 edges, and 6 faces. Now, inserting these
values into Euler's formula, it can be seen that it holds:

$$v - e + f\ = 2$$
$$8 - 12 + 6 = 2$$

Similarly, take the case of the tetrahedron. See Fig. 3.7.

There are 4 vertices, 6 edges, and 4 faces in this case. Thus:

$$v - e + f = 2$$
$$4 - 6 + 4 = 2$$

Euler also demonstrated that for plane figures the value of $v - e + f$ is 1 rather than
2. The Königsberg Bridges graph, being a planar graph, possesses this property. It
has 4 vertices, 7 edges, and 4 faces. See the formula below Fig. 3.9.

Fig. 3.8 Rectangle-plus-
diagonal network

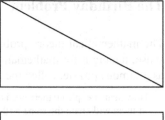

Fig. 3.9 Simple rectangle
network

$$v - e + f = 1$$
$$4 - 7 + 4 = 1$$

Euler proved that this relation was true for any planar figure with a remarkably simple procedure. Take, for example, the rectangle in Fig. 3.8 with a diagonal in it. In terms of Euler's graph theory symbolism, we can see that: $v = 4$, $e = 5$, and $f = 2$:

In this network, there are, as just indicated, 4 vertices, 5 edges (including the diagonal), and two faces—the two triangular segments on either side of the diagonal. The relevant formula is the following one:

$$v - e + f = 1$$
$$4 - 5 + 2 = 1$$

If we remove the diagonal, which is an edge, we also decrease the number of faces by 1. Since the number of vertices remains unchanged, the relation holds. See Fig. 3.9.

$$v - e + f = 1$$
$$4 - 4 + 1 = 1$$

This proves Euler's hypothesis that a plane figure will always show the structure of $v - e + f = 1$, since the figure involved was *any* four-sided plane graphical figure. In general, if we remove an edge from a graph we are simultaneously removing a face from it. This leaves the value of the relation unaltered. If we eliminate a vertex, then we are also removing the edge that goes into it, of course. This reduces "v" and "e" by one. Such modifications leave the formula unchanged.

The aim of the above discussion has been to show how some aspects of mathematical discovery might unfold and how this type of discovery is tied to the cognitive flow model described previously. At first, Euler's insight came from a practical situation that he expressed in the form of a puzzle—an insight that is essentially an abduction. This then led him to extract from his ingenious proof a new way of describing geometrical structure. The end result was the discovery of a previously hidden pattern in geometrical figures, leading to a new branch of mathematics (graph theory).

The Birthday Problem

The mathematical theory probability is a way of quantifying chance and thus, in a sense, taming it for mathematical analysis. Randomness thus becomes less random, as a famous puzzle, called the Birthday Problem, saliently reveals (Ball 1972):

> How many people do there need to be in a room so that there is at least a 50% chance that two of them will share the same birthday?

The origins of the problem are unclear. W. W. Rouse Ball (1972: 60) suggested that it may have been devised by English mathematician Harold Davenport. It came to broad attention in Martin Gardner's column in *Scientific American* in 1957. The Aha insight is to ask the same question for different numbers of people, calculating the relevant probabilities, up till when the probability first drops below 50%. So, let's suppose that there are 2 people in a room. The total number of possible occurrences of birthdays in this case is:

$$365 \times 365$$

If the two people do indeed have different birthdays, then the first one, say A, may have it on any day of the year (365 possibilities) and the second one, say B, may have it on any day except the day of A's birthday. Thus, there are 364 possibilities for B's birthday. The number of possible pairs of distinct birthdays is given by the following formula:

$$365 \times 364$$

The probability of two birthdays falling on the same date is:

$$\frac{364 \times 365}{365 \times 365}$$

Now, without going into the mathematical details, we can generalize this approach to n people, assuming that every single person has a different birthday: A may have it on any of 365 days, B, on any of the remaining 364 days, C, on any of the then remaining 363 days, and so on, until the last, or nth person, who will have their birthday on any of the remaining (365-n) days. The first value for n for which the probability of a pair is below 0.5 is 23. This means that 23 people will do the trick—a truly unexpected finding. Figure 3.10 illustrates the solution visually.

The Birthday Problem is part of a class of probability puzzles that involve permutations and combinations of elements, which takes us right back to the archetype inherent in Alcuin's River-Crossing Puzzle and others. It is also a model of how combinatorial structure and probability factors interact. It is relevant to note that probability theory itself arises from gambling and attempts to understand seemingly random outcomes. The foundations of the modern-day theory were laid by Girolamo Cardano, himself an avid gambler, in the sixteenth century. He was among the first to discuss and calculate the probability of throwing certain numbers

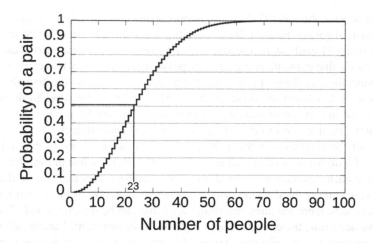

Fig. 3.10 The birthday problem (Wikimedia Commons)

with dice and of pulling certain cards, such as aces, from decks of cards. In his *Book of Games of Chance* (1663), Cardano illustrated how the games he played suggested a mathematical way to represent the outcomes with predictive value.

Galileo also made relevant observations, based on his own game, *passadieci*, which the Grand Duke of Tuscany had asked him to explain. Why is it that by throwing three dice the scores of 10 and 11 occurred more often than 9 and 12? Galileo realized that the numbers 10 and 11 can be obtained with more combinations of dots on the dice, and so they are more likely to come up. Like Cardano, Galileo based his argument on the specification of the elementary events that will occur in an aleatory system, such as card and dice games. In the subsequent century Blaise Pascal, Christian Huygens, Jakob Bernoulli, and Pierre de Fermat developed these ideas into modern probability theory, with Huygens' treatise *De Ratiociniis in Ludo Alea* (1657), constituting the first formal statement of this theory. Huygens stated in the book that seemingly trivial card and dice games harbored a "very interesting and deep theory." In his book *Ars Conjectandi*, Bernoulli articulated the first limit theorem of probability theory—the law of large numbers (Borovkov 2013).

The Birthday Problem is essentially an experiment in probability theory, revealing hidden structure with its counterintuitive answer. It also models how an open puzzle frequently produces a surprise ending, much like a mystery story. Nevertheless, the method of solution behind it is rather straightforward. It is similar to any simple problem in probability, such as the following one, which is discussed briefly here for the sake of argument:

In how many different ways can four aces be drawn blindly from a standard deck?

First, we determine how many ways there are in drawing out any four cards from a 52-card deck. The answer, as in the Birthday Problem, is a factorial:

$52 \times 51 \times 50 \times 49 = 6,497,400$. The reason is that any one of the 52 cards can be drawn first, of course. Each of the 52 possible first cards can be followed by any of the remaining 51 cards, drawn second. Since there are 51 possible second draws for each possible first draw, there are 52×51 possible ways to draw two cards from the deck. Now, for each draw of two cards, there are 50 cards left in the deck that could be drawn third. Altogether, there are $52 \times 51 \times 50$ possible ways to draw three cards. Reasoning the same way, it is obvious that there are $52 \times 51 \times 50 \times 49$, or 6,497,400, possible ways to draw four cards from a standard deck.

To find the chances of getting the aces, it is then necessary to determine the number of four-ace draws there are among the 6,497,400 possible draws. We start by looking at each outcome, draw by draw. There are four aces that can be drawn in some order (among the drawing possibilities). For each one of these, and ignoring the other cards, there are three remaining aces that can be drawn second. Then, for each two-ace draw, there are two aces that can be drawn third. Finally, after three aces have been drawn, only one remains. So, the total number of four-ace arrangements is: $4 \times 3 \times 2 \times 1 = 24$. Thus, among the 6,497,400 ways to draw four cards there are 24 ways to draw the aces. The probability of doing so is, therefore, $24/6,497,400 = 0.0000036$, which makes it a highly unlikely outcome. Although the problem is different from the Birthday Problem, the reasoning behind it is analogous.

As an example of a similar open (rather than closed) puzzle, to which the same mode of reasoning applies, consider the following one devised by Kasner and Newman (1940: 244):

> What is the probability of obtaining either a 6 or a 7 in throwing a pair of dice?

We start by listing the number of ways to throw either a 6 or a 7. See Table 3.2.

Now, there are 36 possible throws of two dice, because each of the 6 faces of the first die is matched with any of the 6 faces of the second one. Of these 36 possible throws, 11 produce either a 6 or a 7 (as the table shows). Therefore the probability of throwing either a 6 or a 7 is 11/36.

As these examples show, probability theory is essentially a means of assigning mathematical pattern to seemingly random events. The pattern that sometimes emerges, as in the case of the Birthday Problem, is unexpected. One of the most famous puzzles in this domain is the so-called St. Petersburg Paradox, which takes its name by an observation put forward by mathematician Daniel Bernoulli.

Table 3.2 Outcomes of dice throwing

Outcome: 6		Outcome: 7	
First die	Second die	First die	Second die
1	5	1	6
2	4	2	5
3	3	3	4
4	2	4	3
5	1	5	2
		6	1

The Paradox was actually devised by his brother Nicolaus Bernoulli (Martin 2004). It can be paraphrased as follows:

A game played at a casino offers a single player a chance to toss a fair coin in stages. Initially, the stake is 2 dollars and this is doubled each time heads appears. The first time tails appears, the game ends and the player wins whatever is in the pot. So, a player will win 2 dollars if tails appears on the first toss, 4 dollars if heads appears on the first toss and tails on the second, eight dollars if heads on the first two tosses and tails on the third, and so on. The player thus wins 2^k dollars (k = number of tosses). What would be a fair price to pay the casino for entering the game?

Answering this question involves figuring out the average payout. The chance of winning 2 dollars is ½ (heads on the first toss); the chance of winning 4 dollars is ¼; the chance of winning 8 dollars is ⅛ ; and so on. We can expect a payoff for each possible outcome of 1 dollar. This can be shown as follows:

$$2 \times \tfrac{1}{2} + 4 \times \tfrac{1}{4} + \ldots + k \times 1/k = 1 + 1 + 1 + \ldots$$

The total expected payoff is clearly an infinite sum. The paradox can be resolved by making a distinction between the amount of the final payoff and the net amount won in the game. When the player first wins he or she will expect the following, according to the stipulations in the statement:

$$2^k - 2 \, \text{dollars}$$

In the toss, the player will have won 2^k dollars. This means that the player will win 2 dollars, no matter how many tosses are taken (Kraitchik 1942: 138–139). Interestingly, the paradox has led to significant debates within probability theory. These need not concern us here. The point is that it brings out the essential characteristics of the cognitive flow theory of puzzles and games, which involves a flow from a situation toward a generalization and a subsequent theoretical engagement. Moreover, the "life lesson" of this paradox is that we do not play games to maximize expected monetary outcomes, but we act as if we were. The mathematics does not lie, providing unconsciously a kind of "deep theory of life," to paraphrase Huygens above.

Paradoxes

The St. Petersburg Paradox leads to the theme of paradoxes within recreational mathematics. Paradoxes have had a significant role in the history of mathematics and logic. One of these is the theory of limits and its derivative, the calculus, whose discovery can be traced ultimately to a set of intellectual memes arising from the paradoxes of Zeno of Elea.

Zeno actually devised some forty paradoxes, but only eight have survived. These were called dialectical by Aristotle, who tried to dismiss them as sophistry, but they resisted his dismissal leading to many new insights into mathematics and logical thinking itself. Zeno's paradoxes concerning motion make up his most famous surviving ones. In one, known as the Dichotomy Paradox, Zeno argued that if we use logical reasoning in an arbitrary way, then we will be forced to conclude that a runner will never reach the end of a race course, even though the runner will actually do so. He "proved" this by stating that the runner first completes half of the course, then half of the remaining distance, and so on infinitely without ever reaching the end. The stages form an infinite series with each term in it half of the previous one:

$$\{1/2 + 1/4 + 1/8 + 1/16 + 1/32 + \ldots\} = \sum_{a}^{b} 1/2^n$$

One could, of course, contend that a paradox such as this one is not a puzzle, strictly speaking. But it can also be argued that its dialectic structure is identical to that of puzzles, concealing a hidden pattern within it that, although seemingly illogical, is nonetheless there. Moreover, the paradox harbors the same kind of infinity archetype which recurs in other puzzles, such as in Ibn Khallikan's chess puzzle and Lucas' Tower of Hanoi puzzle. In addition, it can be claimed that this very type of paradox crystallized as a mathematical meme that eventually led to the theory of limits. In fact, to take some liberty with history, it can be argued that Isaac Newton and Gottfried Wilhelm Leibniz were likely influenced by Zeno's paradox meme when they came up, independently, with an ingenious, yet remarkably simple, solution to it. They simply asserted that the sum to which a series such as $\{1/2, 1/4, 1/8, 1/16 + \ldots + 1/2^n\}$ converges as it approaches infinity is the distance between the starting line and the finish line. Thus, the limit of the runner's movement is, in fact, the unit distance of "1." It is beyond the scope of the present treatment to discuss the historical details behind the origins of the calculus, nor to justify its source in Zeno's paradoxes. Suffice it to say that these likely brought about new views of how we mathematize infinity, suggesting a conceptual framework—limits—for answering some of the classical problems of physics. The theory of limits was not unknown before Newton and Leibniz. As we saw, it was implicit in the proof of π in Ahmes, Archimedes, and others. In effect, the infinity archetype shows up in many and diverse forms.

Paradoxes display the bi-part thinking discussed previously—they both play on logic itself and yet reveal hidden patterns within it. Consider another classic ancient paradox, known generally as the Liar Paradox, which goes back to the fifth century BCE. The story goes that Protagoras, who was the first philosopher to call himself a Sophist, invented the paradox, although its most famous articulation has revolved around a Cretan named Epimenides in the sixth century BCE:

> The Cretan philosopher Epimenides once said: "All Cretans are liars." Did Epimenides speak the truth?

Let's assume that Epimenides spoke the truth. If so, his statement that "all Cretans are liars" is a true statement. However, from this we must deduce that Epimenides, being a Cretan, is thus a liar. But this is a contradiction. Obviously, we must discard our assumption. Let's assume the opposite, namely that Epimenides is in fact a liar.

But, then, if he is a liar, the statement he just made—"All Cretans are liars"—is true. But this is again a contradiction—liars do not make true statements. Obviously, we are confronted with a circularity, not unlike the proverbial chicken and the egg one.

The dream of mathematicians has been to provide a logical foundation to mathematics that would be free of such paradoxical thinking (Davis and Hersh 1986). But paradoxes such as this one have stood in the way of this master plan. They expose logical systems, understood as systems of axioms and proofs that are derived in sequence, as having flaws within them. The source of the circularity in the Liar Paradox is, of course, the fact that it was Epimenides, a Cretan, who made the statement that "all Cretans are liars." It is an example of the logical problems that arise from self-reference. Bertrand Russell found the paradox to be especially troubling, feeling that it threatened the very foundations of mathematics based on propositions that can be proved as true or false and connected to each other. To examine the nature of self-reference more precisely, Russell (1918) formulated his own version of the same paradoxical structure, called the Barber Paradox:

> The village barber shaves all and only those villagers who do not shave themselves. So, shall he shave himself?

Let us assume the barber decides to shave himself. He would end up being shaved, of course, but the person he would have shaved is himself. And that contravenes the requirement that the village barber should shave "all and only those villagers who do not shave themselves." The barber has, in effect, just shaved someone who shaves himself. So, let us assume that the barber decides not to shave himself. But, then, he would end up being an unshaved villager. Again, this goes contrary to the stipulation that he, the barber, shave "all and only those villagers who do not shave themselves"—including himself. It is not possible, therefore, for the barber to decide whether or not to shave himself. Russell argued that such "undecidability" arises because the barber is himself a member of the village. If the barber were from a different village, the paradox would not arise.

Like German philosopher Gottlob Frege (1879), Russell sought to find a system of propositions that would exclude self-reference, circularity, and undecidability. Using a notion developed two millennia earlier by Chrysippus of Soli, Frege claimed that circularity could be avoided from statements such as the Liar Paradox by considering their *form* separately from their *content*. In this way, one could examine the consistency of propositions, without having them correspond to anything in the real world (such as barbers, villages, and Cretans). Frege's approach was developed further by Cambridge logician Ludwig Wittgenstein (1922), who used symbols rather than words to ensure that the form of a proposition could be examined in itself for logical consistency separate from any content to which it could be applied. The problem, Wittgenstein affirmed, was that we expect logic to interpret reality for us. But that is expecting way too much from it. Wittgenstein's system came to be known as "symbolic logic"—a system of representation prefigured actually by Lewis Carroll in his ingenious book *The Game of Logic* (reprinted 1958a).

Russell joined forces with Alfred North Whitehead to formulate a system of symbolic logic, called the *Principia Mathematica* in 1913, in which formal (form-based) propositional structure would obviate any circularity or undecidability that might otherwise arise within it. But it became obvious after publication of the book that the system led to unexpected problems. To solve these, Russell introduced the notion of types, whereby certain types of propositions would be classified into different levels (more and more abstract) and thus considered separately from other lower types. This seemed to avoid the problems—for a while anyhow. The Polish mathematician Alfred Tarski (1933) developed Russell's theory of types further by naming each level of higher abstract statements a *metalanguage*. A metalanguage is a statement about another statement. At the bottom of the metalanguage hierarchy are straightforward statements about things such as "Cretans are liars." Now, if we say that "The statement that Cretans are liars is true," we are using a different type of language, because it constitutes a statement about a previous statement. It is a metalanguage. The problem with this whole approach is that more and more abstract metalanguages are needed to evaluate lower-level statements and to avoid undecidability. But this will go on ad infinitum. So, paradoxically, the system of metalanguages is itself undecidable.

In their 1986 book, *The Liar*, mathematician Jon Barwise and philosopher John Etchemendy adopted a practical view of the Liar Paradox and of paradoxes generally. As they assert, the paradox arises only because we allow it to arise, indirectly supporting the assertion here that paradoxes belong to the realm of ludic thinking. When Epimenides says, "All Cretans are liars," he may be doing so simply to confound his interlocutors. His statement may also be the result of a slip of the tongue. Whatever the case, the intent of Epimenides' statement can only be determined by assessing the context in which it was uttered along with Epimenides' reasons for saying it. Once such factors are determined, no paradox arises. Nevertheless, the paradox has had enormous implications for the very concept of propositional logic.

This whole line of reasoning was brought to an end in 1931 by Kurt Gödel, who showed why undecidability is a fact of the human brain, no matter how hard we try to eliminate it from our logical systems. Before Gödel, it was taken for granted that every proposition within a mathematical system could be either proved or disproved within that system. This is exactly what Euclid did in his *Elements*. But Gödel startled the mathematical world by showing that this was not the case. He argued that a logical system invariably contains a proposition within it that is "true" but "unprovable." Gödel's argument is far too technical to be taken up in an in-depth manner here. For the present purposes, it can be paraphrased as follows (Danesi 2002):

> Consider a mathematical system, T, that is both correct—in the sense that no false statement is provable in it—and contains a statement "S" that asserts its own unprovability in the system. S can be formulated simply as: "I am not provable in system T." What is the truth status of S? If it is false, then its opposite is true, which means that S is provable in system T. But this goes contrary to our assumption that no false statement is provable in the system. Therefore, we conclude that S must be true, from which it follows that S is unprovable in T, as S asserts. Thus, either way, S is true, but not provable in the system.

Raymond Smullyan (1997) provides a clever puzzle version of Gödel's argument as follows:

Let us define a logician to be accurate if everything he can prove is true; he never proves anything false. One day, an accurate logician visited the Island of Knights and Knaves, in which each inhabitant is either a knight or a knave, and knights make only true statements and knaves make only false ones. The logician met a native who made a statement from which it follows that the native must be a knight, but the logician can never prove that he is! What was the statement?

The "Gödelian" statement is: *You cannot prove that I am a knight.* If the native were uttered by a mendacious knave, then the statement would, of course, be false. Its opposite would then be true—namely, *You can prove that I am a knight.* But according to our assumption the native is not a knight; he is a knave. Since the puzzle asserts that an accurate logician is incapable of proving anything false, he cannot prove the uttered falsehood. So, let's assume the native is a knight. This means that the statement *You cannot prove that I am a knight* is true. But if it is true, then the logician cannot prove it—the statement declares as much. So, even though the native is a knight, the logician will never be able to prove it. Gödel's overall demonstration showed, like Smullyan's puzzle, that there will always be a proposition in a system that is undecidable—it cannot be proved to be either true or false. Alan Turing's brilliant 1936 paper followed by arguing that some objects cannot be computed, which is another way of saying that they are undecidable. An undecidable problem in computer science is one for which it is impossible to construct a single algorithm that always leads to a correct yes-or-no answer.

It can be suggested that all paradoxes contain an intrinsic ludic archetype that is similar to Jung's trickster archetype—inhering in a tricky entanglement of opposites. For every logical aspect of human thought, there is an equal and opposite trickster thought. The ludic archetype is a breaker of logical assurance, providing comic relief but at the same time teaching a moral lesson or exposing the folly of human conceit. A classic trickster paradox is the Unexpected Hanging Paradox. It can be paraphrased as follows (Margalit and Bar-Hillel 1983, Shapiro 1998):

A condemned logician is to be hanged at noon, between Monday and Friday. But he is not told which day it would be. As he waits, the logician reasons as follows: "Friday is the final day available for my hanging. So, if I am alive on Thursday evening, then I can be certain that the hanging will be Friday. But since the day is unexpected, I can rule that out, because it is impossible. So, Friday is out. Thus, the last possible day for the hanging to take place is Thursday. But, if I am here on Wednesday evening, then the hanging must perforce take place on Thursday. Again, this conflicts with the unexpectedness criterion of the hanging. So, Thursday is also out." Repeating the same argument, the logician is able to rule out the remaining days. The logician feels satisfied, logically speaking. But on Tuesday morning he is hanged, unexpectedly as had been promised.

This is a truly clever demonstration of how one can reason about anything, and yet how the reasoning might have nothing to do with reality. Are formalist theories, such as the ones put forward by Frege and Russell, subject to the Unexpected Hanging paradox? Aware of the profoundly disturbing aspect of trickster paradoxes, David Hilbert (1931) proposed a set of requirements that a logical theory of mathematics should observe. Known as *Hilbert's program*, it was written just before

Gödel's theorem as a framework for rescuing mathematics from the trickster archetype in cognition. Hilbert's program included the following criteria:

1. *Formalization:* a complete formalization of mathematics, with all statements
   articulated in a precise formal language that obeyed well-defined rules.
2. *Completeness:* the formal system must show that all mathematical statements
   are true.
3. *Consistency:* there must be a proof that no contradiction can be obtained in the
   formal set of rules.
4. *Conservation:* there must be a proof that any result relating to "real things" by
   using reasoning about "ideal objects" can be provided without the latter.
5. *Decidability:* an algorithm must be determined for deciding the truth or falsity of
   any mathematical statement.

Current versions of mathematical logic, proof theory, and so-called "reverse"
mathematics, are based on realizing Hilbert's program—reverse mathematics is a
system that seeks to establish which axioms are required to prove mathematical
theorems, thus turning the Euclidean system of proof upside down, going in reverse
from the theorems to the axioms, rather than from the axioms to the theorems.

Hilbert's program was based on the hope that mathematics could be formalized
into one system of the predicate calculus (propositional structure), whether or not it
linked mathematics to reality. This is known as *logicism*—the attempt to make logic
the core of mathematics and then to connect it to reality. Aware of the issues
connected with this stance, Hilbert made the following insightful statement (cited
in Tall 2013: 245):

> Surely the first and oldest problems in every branch of mathematics spring from experience
> and are suggested by the world of external phenomena. Even the rules of calculation with
> integers must have been discovered in this fashion in a lower stage of human civilization,
> just as the child of today learns the application of these laws by empirical methods. But, in
> the further development of a branch of mathematics, the human mind, encouraged by the
> success of its solutions, becomes conscious of its independence. It evolves from itself alone,
> often without appreciable influence from without, by means of logical combination, gener-
> alization, specialization, by separating and collecting ideas in fortunate ways, in new and
> fruitful problems, and appears then itself as the real questioner.

Without going here into the many responses to Hilbert's program, including the
$P = NP$ problem, it is sufficient to point out that none of this line of inquiry would
have been likely possible without trickster paradoxical archetypes. As Tall (2013:
246) comments, mathematicians must simply lower their sights, continuing to use
formalism only when and where it is applicable:

> Instead of trying to prove *all* theorems in an axiomatic system (which Gödel showed is not
> possible), professional mathematicians continue to use a formal presentation of mathematics
> to specify and prove many theorems that are amenable to the formalist paradigm.

Incidentally, it is relevant to note that the very concept of "logic" itself does not
originate in the world of mathematics but in a more mystical domain. It was in sixth-
century BCE Greece that the philosopher Heraclitus asserted that the world was
governed by the *Logos*, a divine force that produces order in the flux of Nature.

Logos came to be viewed shortly thereafter as a rational divine power that directed the universe. Through the faculty of reason, all human beings were thought to share in it. Even the Gospel according to John identifies *Logos* ("the Word") as a spiritual force: "In the beginning was the Word, and the Word was with God."

The above discussion suggests an "anthropic principle," which implies that we are part of the world in which we live and thus privileged, in a way, to understand it best. Al-Khalili (2012: 218) puts it as follows:

> The anthropic principle seems to be saying that our very existence determines certain properties of the Universe, because if they were any different we would not be here to question them.

So, we both invent and discover mathematics, but some mathematics, wherever it is hidden as an obscure secret (to refer back again to Ahmes) may not be "discoverable." We could conceivably live in a world without the Pythagorean theorem, for example. But our bi-part mind wants to turn intuition into ratiocination, as discussed several times already. So, we have devised the theorem to explain to our logical part of the brain what we know intuitively—that a diagonal distance is shorter than taking an L-shaped path to a given point. And perhaps this is why it emerged—it suggests that we seek to understand why we do things intuitively.

Epimenides' paradox has fascinated logicians and philosophers throughout history for this very reason—it seems to imply something that is just beyond our mental reach to understand. As Eugene Northrop (1944: 12) aptly puts it:

> The case of the self-contradicting liar is but one of a whole string of logical paradoxes of considerable importance. Invented by the early Greek philosophers, who used them chiefly to confuse their opponents in debate, they have in more recent times served to bring about revolutionary changes in ideas concerning the nature of mathematics. For some reason, paradoxical statements have a bizarre appeal to all but diehard Aristotelian logicians.

Paradoxes are especially alluring to children for the reason that they stimulate the ludic (trickster) part of the brain, as Henry Dudeney (1958: 15) perceptively observed:

> A child asked, "Can God do everything?" On receiving an affirmative reply, she at once said: "Then can He make a stone so heavy that He can't lift it?"

The child's question is similar to a philosophical conundrum: *What would happen if an irresistible moving body came into contact with an immovable body?* As Dudeney observes, such bizarre paradoxes arise only because we take delight in inventing them. In actual fact, "if there existed such a thing as an immovable body, there could not at the same time exist a moving body that nothing could resist." In the 1960s, an attempt was made by Berkeley professor Lofti Zadeh (1965) to incorporate the real-life value of statements directly into logic, rather than eliminate them *à la* Frege and Wittgenstein. Zadeh proposed a kind of "fuzzy logic" (versus propositional logic) that would envision such matters as Epimenides' statement as a "half truth" or a "half falsehood," depending on context, and the barber's statement as "true under some conditions," but "false under others." But fuzzy logic has hardly solved the dilemma of the liar paradox. It has dodged it in a clever manner by bringing "real life" into the picture. But this raises a series of related paradoxical questions: What is real life? Why do people lie? A more useful approach would be simply to outlaw formulas, propositions, or procedures within logical systems that lead to inconsistencies or circularities. Such an approach has, in fact, already been taken by mathematicians in prohibiting division by 0. Permitting that operation would lead to contradictory results. But this prohibition has hardly stopped mathematicians from doing mathematics. On the contrary, it has led to increasingly new vistas about the nature and referential domain of numbers.

Paradoxes warn us that in generalizing from specific facts we run into difficulties. As Kasner and Newman (1940: 219) aptly put it: "In the transition from *one* to *all*, from the specific to the general, mathematics has made its greatest progress, and suffered its most serious setbacks, of which the logical paradoxes constitute the most important part." The belief that science and mathematics would be able to explain all mysteries, miracles, and hidden patterns, and therefore that the past and the future could be contained, developed in the post-Galilean era of intellectual history. But nothing has proven to be farther from the truth. As products of the dialectic mind, paradoxes warn us to this day against complacency about our own mental creations. Like a court jester, the *homo ludens* in our brain is always prepared to contrive some paradox that will mischievously undermine our most elaborate logical concoctions. As Rucker (1987: 218) aptly observes, the "great dream of rationalism has always been to find some ultimate theory that can explain *everything*." Perhaps what makes paradoxes so "mischievously appealing" is that they reveal ultimately why a "theory of everything" is impossible. They emphasize that human systems are ultimately "imperfect." Gödel's theorem showed, in effect, that logic was made by imperfect logicians, and thus that the dream of using logic to solve all human problems is illusory. Nevertheless, buried deeply within the imagination is the belief that if we were to solve all the puzzles and paradoxes that come from the world we could effect change in that world, because we would have discovered the mystical structures that it conceals—Ahmes' "obscure secrets."

Discovery

The gist of the above discussion is that discovery in mathematics is rarely spontaneous and disconnected with puzzles and paradoxes. As argued throughout this book, the spread of puzzle memes is triggered if there is archetypal structure present in the original puzzle format. Given this condition, the discovery process is then set in motion, leading to complex ideas and, in some cases, new branches of mathematics.

Needless to say, not all discoveries occur in this way. There are many varied factors involved in how mathematics develops. But the overall process seems to be a consistent one: first, there is some idea that comes into view through some format (such as puzzles); second, this is transmitted unconsciously to mathematicians who typically generalize it through some formal algebraic (or other notational) system;

and third, it then becomes known generally in the mathematics community, embedding itself into new systems of thought and methodology.

Archetypal thinking is a basis of discovery in mathematics. A puzzle may possess more than one archetype—a phenomenon that occurs more often than not. The Rabbit Puzzle, for instance, contains both the infinity and recursion archetypes; Alcuin's River-Crossing Puzzles, Euler's Königsberg Bridges Puzzle, and Lucas' Tower of Hanoi enfold arrangement and movement archetypes that led to critical path analysis, graph theory, topology, and the like. Probability puzzles and paradoxes conceal archetypes of possibility and impossibility. In effect, puzzles are imaginative formats that embed archetypal ideas as they originated in the mind of the mathematician. One more episode in the history of mathematics bears this out and is worth revisiting here, since it concerns the archetype of infinity again, which is often the source of remarkable paradoxes. In the sixteenth century Galileo noticed a paradoxical pattern—if the counting numbers, $\{1, 2, 3, 4, \ldots\}$, are compared, one-by-one, to the numbers in one of its subsets in sequence, such as the square numbers $\{1, 4, 9, 16, \ldots\}$, something astounding happens. See Fig. 3.11.

No matter how far one continues along this layout of numbers, there will never be a gap between the top line and the bottom one. This suggested to Galileo that the "number" of elements in the set of all positive integers and the "number" of those in one of its proper subsets is the same. If one stops to think about it, this is indeed an astonishing paradox. To this day, we have a hard time grasping its implications. As Clark (2012: 74) aptly puts it: "We are so accustomed to thinking of finite collections that our intuitions become disturbed when we first consider infinite sets like the set of positive integers."

It is instructive to note that such seemingly "disturbing" results, as Clark characterizes them, derived from simply using a diagrammatic line representation of the two sets of numbers and putting them in a one-to-one relation. In other words, it was the structure of the layout that suggested the pairing technique and produced the astonishing result. It constitutes a discovery that falls easily within the realm of Ahmes' obscure secrets. This truly mysterious happenstance discovery no doubt inspired Georg Cantor's (1874) incredible demonstrations about infinite sets. Like Galileo he showed that the set of counting numbers, also called cardinal numbers, can be matched against any of its subsets, such as the even numbers. See Fig. 3.12.

1	2	3	4	5	6	7	8	9	10	11	12	13	14	...
↕	↕	↕	↕	↕	↕	↕	↕	↕	↕	↕	↕	↕	↕	...
1	4	9	16	25	36	49	64	81	100	121	144	169	196	...

Fig. 3.11 Galileo's integer-to-square integer correspondence

1	2	3	4	5	6	7	8	9	10	11	12	13	14	...
↕	↕	↕	↕	↕	↕	↕	↕	↕	↕	↕	↕	↕	↕	...
2	4	6	8	10	12	14	16	18	20	22	24	26	28	...

Fig. 3.12 Integer-to-even integer correspondence

Cantor pointed out that such sets have the same "cardinality," or same number of elements that can be counted with the cardinal numbers. Incredibly, from this kind of demonstration have come many insights into the nature of numbers, sets, and mathematical infinity. Galileo suspected that mathematical infinity posed a serious challenge to common sense. It was in his 1632 *Dialogue Concerning the Two Chief World Systems* that he noted that the set of square integers can be compared with all the whole numbers (above), leading to the preposterous possibility that there may be as many square integers as there are numbers (even though the squares are them-selves only a part of the set of integers). Now, what is even more bizarre is the fact that the same one-to-one correspondence can be set up between the whole numbers and numbers raised to any power. See Fig. 3.13.

Cantor's demonstrations were earth shattering in mathematical circles when he first made them public. But his take on the infinity archetype did not stop there. It led to more astounding discoveries within infinite sets, cumulatively leading to the modern theory of sets.

Consider the set of rational numbers. These are numbers that can be written in the form p/q where p and q are integers (and $q \neq 0$). The cardinal numbers are, themselves, a subset of the rational numbers—every integer p can clearly be written in the form $p/1$. Terminating decimal numbers are also rational, because a number such as 3.579 can be written in p/q form as 3579/1000. Finally, all repeating decimal numbers are rational, although the proof of this is beyond the scope of the present discussion. For example, 0.3333333... can be written as 1/3. Amazingly, Cantor demonstrated that the rational numbers have the same cardinality of the counting

1	2	3	4	5	6	7	8	9	10	11	12	...
$\updownarrow$	$\updownarrow$	$\updownarrow$	$\updownarrow$	$\updownarrow$	$\updownarrow$	$\updownarrow$	$\updownarrow$	$\updownarrow$	$\updownarrow$	$\updownarrow$	$\updownarrow$	
1^n	2^n	3^n	4^n	5^n	6^n	7^n	8^n	9^n	10^n	10^n	12^n	...

Fig. 3.13 Integer-to-exponential integer correspondence

Fig. 3.14 Cantor's diagonal proof (Wikimedia Commons)

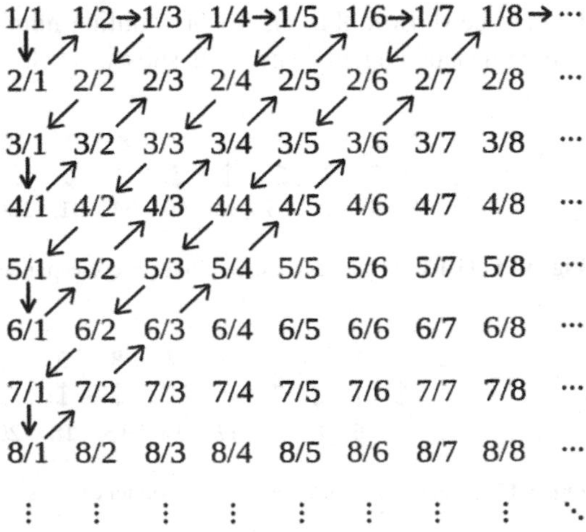

$$
\begin{array}{cccccccc}
1/1 & 1/2 \rightarrow 1/3 & 1/4 \rightarrow 1/5 & 1/6 \rightarrow 1/7 & 1/8 \rightarrow \cdots \\
2/1 & 2/2 & 2/3 & 2/4 & 2/5 & 2/6 & 2/7 & 2/8 & \cdots \\
3/1 & 3/2 & 3/3 & 3/4 & 3/5 & 3/6 & 3/7 & 3/8 & \cdots \\
4/1 & 4/2 & 4/3 & 4/4 & 4/5 & 4/6 & 4/7 & 4/8 & \cdots \\
5/1 & 5/2 & 5/3 & 5/4 & 5/5 & 5/6 & 5/7 & 5/8 & \cdots \\
6/1 & 6/2 & 6/3 & 6/4 & 6/5 & 6/6 & 6/7 & 6/8 & \cdots \\
7/1 & 7/2 & 7/3 & 7/4 & 7/5 & 7/6 & 7/7 & 7/8 & \cdots \\
8/1 & 8/2 & 8/3 & 8/4 & 8/5 & 8/6 & 8/7 & 8/8 & \cdots \\
\vdots & \vdots & \vdots & \vdots & \vdots & \vdots & \vdots & \vdots & \ddots
\end{array}
$$

1	2	3	4	5	6	7	8	9	10	...
↕	↕	↕	↕	↕	↕	↕	↕	↕	↕	
1/1	2/1	1/2	1/3	3/1	4/1	3/2	2/3	1/4	1/5	...

Fig. 3.15 Cantor's correspondence between the integers and the rationals

numbers. His method of proof is, again, unexpectedly elegant and simple. First, he arranged the set of all rational numbers as shown in the array in Fig. 3.14.

In each row the successive denominators (q) are the integers {1, 2, 3, 4, 5, 6, 7, ...}. The numerator (p) of all the numbers in the first row is 1, of all those in the second row 2, of all those in the third row 3, and so on. In this way, all numbers of the form p/q are covered in the above array. Cantor highlighted every fraction in which the numerator and the denominator have a common factor. If these fractions are deleted, then every rational number appears once and only once in the array. Now, Cantor set up a one-to-one correspondence between the integers and the numbers in the array as follows: he let the cardinal number 1 correspond to 1/1 at the top left-hand corner of the array; 2 to the number below (2/1); following the arrow, he let 3 correspond to 1/2; following the arrow, he let 4 correspond to 1/3; and so on, ad infinitum. The path indicated by the arrows, therefore, allows us to set up a one-to-one correspondence between the cardinal numbers and all the rational numbers. See Fig. 3.15.

This layout shows that there are as many rational numbers as there are whole numbers. One cannot help but be impressed by the simple way in which Cantor constructed this mind-boggling demonstration. In effect, once the simplicity inherent in the principles of Cantor's overall theory is understood, they cease to look like the products of the overactive imagination of a mathematical eccentric. Cantor then classified those numbers with the same cardinality as belonging to the set "aleph null," or $\aleph_0$ ($\aleph$ is the first letter of the Hebrew alphabet). He called $\aleph_0$ a *transfinite number*. Amazingly, Cantor discovered that there are other transfinite numbers. These are sets of numbers with a greater cardinality than the integers. He labeled each successively larger transfinite number with increasing subscripts {$\aleph_0, \aleph_1, \aleph_2, \ldots$}.

Cantor's proof is again remarkable for its simplicity. Suppose we take all the possible numbers that exist between 0 and 1 on the number line and lay them out in decimal form. Let's label each number {$N_1, N_2, \ldots$}. We note that there are so many possible numbers of the form p/q between 0 and 1 that we could not possibly put them in any order. So, the numbers given here are just a sampling.

$$N_1 = 0.4225896\ldots$$
$$N_2 = 0.7166932\ldots$$
$$N_3 = 0.7796419\ldots$$
$$\ldots$$

How could we possibly construct a number that is not on that list? Let's call it C. To create it, we do the following: (1) for its first digit after the decimal point we choose a number that is greater by one than the first digit in the first place of N_1;

(2) for its second digit we choose a number that is greater by one than the second number in the second place of N_2; (3) for its third digit we choose a number that is greater by one than the third number in the third place of N_3; and so on:

$N_1 = 4225896\ldots$

The constructed number, C, would start with 5 rather than 4 after the decimal:

$C = 0.5\ldots$

$N_2 = 0.7166932\ldots$

The constructed number would have 2 rather than 1:

$C = 0.52\ldots$

$N_3 = 0.7796419\ldots$

The constructed number would have 0 rather than 9

$C = 0.520\ldots$

$\ldots$

Now, the number $C = 0.520\ldots$ is different from $N_1, N_2, N_3, \ldots$ because its first digit is different from the first digit in N_1; its second digit is different from the second digit in N_2; its third digit is different from the third digit in N_3, and so on ad infinitum. We have in fact just constructed a different transfinite number than $\aleph_0$. It appears nowhere in the list above.

The gist of the foregoing discussion and illustrations is that all this would not have occurred without a hunch—laying out numbers in some correspondence up through infinity. The same infinity archetype is at play here as it was in Zeno and other paradoxes and puzzles. Of course, by making such blanket statements, one is belaboring the obvious; that is, by asserting that something is an archetype may be just a semantic reformulation. Nonetheless, had the brain had a different structure, archetypes such as infinity and the puzzles and paradoxes it has generated would likely not have come into existence.

Of course, a way around the brain-as-mind-as-brain vexatious circularity is to eliminate the distinction between inner (mental) and observable (behavioral) processes and to create artificial models of cognition that will become self-sustaining (Kurzweil 2012). While this seems to be a modern premise, it really is no more than a contemporary version of the ancient belief that the human mind is a machine programmed to receive and produce information in biologically determined ways. In a way, the search for understanding paradoxes and paradoxical proofs such as those by Cantor are a consequence of consciousness. Daniel Bor (2012) describes consciousness as pattern-seeking behavior. A slight modification to this character-ization is that consciousness is the end-product of a flow from the collective unconscious, where archetypes reside, to states of conscious understanding.

Chapter 4
Puzzles and Spatial Reasoning

> *Mathematics would certainly have not come into existence if*
> *one had known from the beginning that there was in nature no*
> *exactly straight line, no actual circle, no absolute magnitude.*
>
> —Friedrich Nietzsche (1844–1900)

A main tenet of this book is that the idea-structure for many puzzles typically originates in the imagination (a right-hemispheric function) and then migrates, via a cognitive flow, to embed itself into the reasoning and generalizing part of the brain (a left-hemispheric function). The imagination defies a precise definition; that is, we cannot pin it down to a clear semantic explication. One of its most basic functions is to conjure up images in the mind and connect them in some meaningful way to allow people to accomplish particular tasks, such as solving problems in geometry. This primary level of imagination now comes under the rubric of "spatial reasoning," a term that can be traced to Howard Gardner's work on what he called "multiple intelligences" in 1983, in contrast to the idea of a single general form of intelligence. Gardner identified eight intelligences:

1. Musical-rhythmic
2. Visual-spatial
3. Verbal-linguistic
4. Logical-mathematical
5. Bodily or kinesthetic
6. Interpersonal
7. Intrapersonal
8. Naturalistic

With this model, Gardner wanted to contravene the notion of a general intelligence, embracing the possibility of non-general modes of intelligence that are connected with bodily experiences and social skills. However, for some psychologists, Gardner's taxonomy turns out to be indistinguishable from the traditional view that there is an overall intelligence, with subcategories that are more or less in line with the multiple intelligences of which Gardner speaks. A study by Visser, Ashton, and Vernon (2006), for instance, argued that Gardner's model did not end up

© Springer International Publishing AG, part of Springer Nature 2018   105
M. Danesi, *Ahmes' Legacy*, Mathematics in Mind,
https://doi.org/10.1007/978-3-319-93254-5_4

contravening the notion of general intelligence, at least as a construct for investigating cognitive abilities, but actually confirming the notion of a general intelligence with variant manifestations. For the present purposes, there is no specific locus within Gardner's taxonomy for what has been called here dialectic-ludic thinking. What is of relevance to the present discussion, however, is that Gardner identifies visual-spatial intelligence, or spatial reasoning, as a factor in cognition. This notion has, in fact, been employed practically in math education and psychology, and it can be enlisted here as well to discuss how primary imaginative thinking in visual puzzle formats unfolds and how it is connected to mental imagery. For the present purposes, therefore, spatial reasoning can be defined as the ability of the imagination to grasp the meaning of spatial relations among given objects and figures, including mental folding and rotation, arrangements of forms, and so on.

The topic of mental imagery has a long history in psychology. The relevant research has shown that images can be elicited specifically to carry out various thinking tasks. People can imagine faces and voices accurately and quickly, rotate objects in their heads, locate imaginary places, scan game boards (like a checker board) in their minds, and so on with no difficulty whatsoever. Stephen Kosslyn (1983, 1994), who is well known for having investigated empirically how the brain's imagery system might work, conducted a series of ingenious experiments that show how people can easily form images in their mind to help them examine tasks hypothetically, such as arranging furniture in a room, designing a blueprint, and so on, before actually carrying them out. Puzzles and games that involve spatial reasoning constitute the subject matter of this chapter.

Tangrams

As we saw in the opening chapter, Archimedes' *loculus* was devised as an assembly game in which a square is cut into fourteen geometric pieces. The objective is to assemble the scrambled pieces to form silhouettes of objects or else to reconstruct the original square outline. As such, it is a model of spatial reasoning based on a general archetype—assembly of the parts into wholes. Archimedes' game is lost; the version that we have today comes down to us from an Arabic manuscript titled *The Book of Archimedes on the Division of the Figure Stomaschion* (Netz and Noel 2007).

The *loculus* has become a staple within recreational mathematics, because of its mathematical properties that need not concern us here. A descendant of the game, as also briefly mentioned in Chap. 1, is the *tangram*, which has been used as a means to explore the psychology of spatial reasoning itself (Bohning and Althouse 1997). For example, Ayaz, Izzetoglu, Sheworkis, and Onaral (2012) presented evidence that shows how spatial thinking involved in tangrams originates in the visual right hemisphere becoming gradually more abstract—thus providing indirect evidence for the cognitive flow model being proposed in this book.

The origin of the tangram is unknown. Most believe that it originated in China as a mystical artifact, akin to the magic square (Vorderman 1996: 130). As Slocum and

Fig. 4.1 The Chinese tangram (Wikimedia Commons)

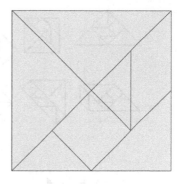

Botermans (1994: 8) suggest, the most likely source of the tangram is the late eighteenth-century Chinese puzzle Ch'i Ch'io, invented during the reign of emperor Chia Ch'ing (1796–1820). But Takagi (1999) argues instead that the tangram may be of Japanese origin because he discovered a book containing seven-piece tangram puzzles, titled *The Ingenious Pieces of Sei Shonagon*, that was published in 1742 in Japan. Actually, there are two tangram configurations (seven pieces with small differences), now known, respectively, as Chinese and Japanese, indicating that they are serendipitous manifestations of the same archetypal structure that can be seen as well in the *loculus*. In this chapter, only the Chinese version will be used. Whatever the real origin of the tangram, the puzzle made its way to Europe and America in the early nineteenth century, where it became very popular. It is said that even Napoleon was an avid tangram player while in exile on St. Helena. Lewis Carroll, Sam Loyd, and Henry Dudeney, among other puzzlists, created ingenious tangram puzzles. Loyd even devoted an entire book to this genre, *The Eighth Book of Tan*, showing his readers how truly intriguing it was (Loyd 1952).

As mentioned, there are seven tangram pieces—five triangles, one square, and one parallelogram. These are cut in such a way that they will fit together to form a large square. See Fig. 4.1.

All the pieces have reflection and rotation symmetry, except for the parallelogram that has only the latter. Its mirror image is obtained by flipping it over. The objective of the game is to assemble these pieces (called tans) in order to produce recognizable shapes, figures, and forms; they can be used as many times as needed to create the required forms (Read 1965). For example, below are thirteen convex shapes made with the seven tans, as well as a set of figures that have recognizable outline shapes, including the outline of a cat and of a house. See Fig. 4.2.

While these may seem on the surface fairly simple to construct, as it turns out, tangrams are very difficult to solve, requiring a large dose of spatial reasoning. As Loyd showed in his tangram puzzle collection, the tangram can also be used to create various intriguing derivative puzzles and paradoxes that lead to spatial-imagistic explorations of hidden ideas and structures. One of these is the following (Loyd 1952). It is paraphrased below. See Fig. 4.3.

> The tangram figures represent the mysterious square, built with seven pieces: then with a corner clipped off, the same seven pieces are employed. How is this so?

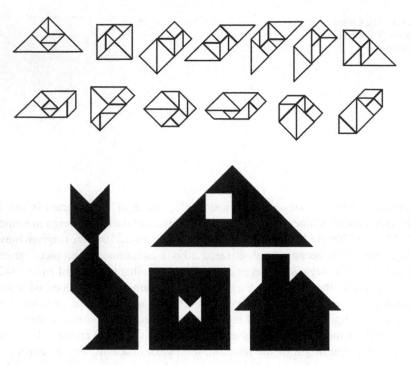

Fig. 4.2 Shapes made with the tans (Wikimedia Commons)

Fig. 4.3 Loyd's tangram
paradox (from Loyd 1952)

This paradox involves a crafty assembly move, whereby one of the pieces is shifted to create the optical illusion. Loyd used a similar ploy with his *Get Off the Earth* puzzle, which will be discussed below. A similar optical illusion paradox, known as the Two Monks Paradox, was devised by Henry Dudeney (1917). It consists of two similar shapes, resembling "monks," which show a missing foot on one of them, even though the same number of pieces were used. See Fig. 4.4.

Like Loyd's puzzle, this is an assembly trick that has a simple explanation—the area of the foot is actually compensated by the slightly larger body in the other figure. As these show, tangrams, like the *loculus,* have an intuitive appeal for the plausible reason that the assembly archetype stimulates the spatial imagination to create pattern from scattered pieces. The same archetype can be seen in perhaps the most well known of all assembly puzzles—the jigsaw puzzle. It was invented by British mapmaker John Spilsbury around 1760 as a toy to educate children about geography (Hannas 1972, 1981, Williams 2004). Jigsaw puzzles for adults were put

Fig. 4.4 Dudeney's two
monks paradox (Wikimedia
Commons)

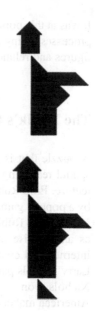

on the market around 1900. Most of these had no guide or picture on the box (unlike children's versions of the puzzles), thus requiring a large dose of spatial reasoning. The Parker Brothers Company introduced picture guides and interlocking pieces in 1908. This new form of the puzzle became so successful that in 1909 the company devoted its entire factory production to it. Made of wood, the puzzles were extremely expensive—a 500-piece puzzle cost 5 dollars in 1908, when the average wage was 50 dollars per month. By 1933, most were made of cardboard, propelling sales to nearly ten million per week in the United States alone. Retail stores offered free puzzles with the purchase of certain gadgets, and a twenty-five-cent magazine called *Jigsaw Puzzle of the Week* appeared on newsstands every Wednesday. Today, the jigsaw puzzle remains one of the more popular types of games. Specialty stores throughout North America sell jigsaw puzzles to suit all tastes.

One could claim, with some creative license, that the entire science of geometry is based on the same kind of archetype, or on interacting spatial archetypes designed to impart assembly structure on different geometric forms. This may be the reason why geometric ideas are easily acquired by children, irrespective of culture. In the late 1950s, Pierre van Hiele and Dieke van Hiele-Geldof noticed that students learning basic geometry tended to do so by first recognizing certain geometric shapes as consistent, regardless of size, orientation, and texture (see van Hiele 1984). In other words, children instinctively recognized the difference between structure and form (a triangle has the same structure, regardless of the particular form it takes). The learners then started to focus on the component parts of figures (sides of a triangle, diameter of a circle, etc.) and their relations to each other (equality, inequality, etc.). This entailed learning to focus on the properties of geometric forms and assemblages and what they stand for. From this phase, they started to understand relations among figures recognizing, for example, that squares and rectangles are quadrilaterals.

It was at this point that students started to work consciously with proofs and abstract processes, being able to formulate general axioms about the nature of geometrical figures and relations on their own.

The Rubik's Cube

A puzzle that is now also used in psychology and math education as a model of spatial reasoning is the Rubik's Cube, invented by a Hungarian professor of architecture, Ernö Rubik, in 1975. It seems that he was inspired (perhaps unconsciously) by a popular game called the Thirty Colored Cubes Puzzle, invented in 1921 by P. A. MacMahon. Rubik taught at the Budapest School of Applied Arts, creating the cube as an exercise in spatial reasoning for his students. By 1982 it had become an international craze, with ten million sold in Hungary alone. An individual named Larry Nichols patented a similar puzzle cube in 1957 (Costello 1988: 148). In 1984, Nichols won a patent infringement lawsuit against the Ideal Toy Company, the American maker of the Rubik's Cube. Nevertheless, Rubik's specific cube, more so than the others, has some unique properties that make it useful for mathematical analysis, including implications for graph theory and topology.

The cube is made up of smaller, colored cubes, so that each of the six faces of the large cube is a different color. The colors can be scrambled by twisting sections of the cube around any axis. The goal is to return the cube to its initial configuration. See Fig. 4.5.

Solutions typically consist of a sequence of spatially imagined moves. For example, one move might entail switching the locations of three corner pieces, while leaving the rest of the pieces in their places. Much research has been done on optimal solutions to the puzzle. The original cube is a $3 \times 3 \times 3$ device in which the subcubes on its outside are connected in such a way that rotation is possible in any plane. Each of the six sides is painted a distinct color, and (to reiterate) the goal of the game is to return the cube to a state in which each side has a single color after it has been randomized. The number of possible permutations involved in solving the cube, considered as a graphic network, is determined as follows (Frey and Singmaster 1982, Rubik 1982). The cube has eight corners and twelve edges. There are thus 8! ways to

Fig. 4.5 Rubik's Cube
(Wikimedia Commons)

arrange these eight corner cubes. Each one has three orientations, even though only 7 of the 8 can be oriented independently. Orienting the eighth corner relies on the other seven, involving therefore 3^7 possibilities. There are 12!/2 ways to arrange the edges, because edges must be in an even arrangement. Eleven edges can be turned independently, with the twelfth depending on the previous ones, producing 2^{11} possibilities. Overall, the number of possible arrangements is:

$$8! \times 3^7 \times (12!/2) \times 2^{11} = 43,252,003,274,489,856,000$$

It has been known since 1995 that a lower bound on the number of moves for the solution (in the worst case) is 20, and it is now known that no configuration requires more than 20 moves. Various other mathematical implications have been derived from this game, including the fact that it manifests group structure. Clearly, the Rubik's Cube is based on the same genus of assembly archetype that undergirds other spatial puzzles—a feature that seems to come up invariably in the analysis of manual-spatial dexterity games. This may, again, be the reason why such games appear in different versions throughout the world and across time.

Flatland

A key psychological principle of the tangram and Rubik's Cube is that they constitute miniature models of how we likely imagine physical objects and situations through ludic forms. As mentioned, geometry itself may be a model of the brain's spatial imagination—a possibility that was brought out in a famous novel with mathematical implications, called *Flatland: A Romance of Many Dimensions*, written by preacher and literary critic Edwin A. Abbott in 1884. It has become famous among recreational mathematicians (and mathematicians generally) because it models how n-dimensional geometric thinking unfolds. The characters of the novel are geometrical figures living in a two-dimensional universe called Flatland. Flatlanders see each other edge-on, and thus as dots or lines, even though, from the vantage point of an observer in three-dimensional space looking down upon them, they are actually lines, circles, squares, triangles, and other geometric figures. To grasp the difference that the viewing perspective makes, it is useful to think of Flatland as the surface of a flat and smooth table. If one crouches to look at a circular piece of paper lying on a table, keeping the eyes level with the table's surface, it will appear as a line. The only way to see it as a circle is to view it from above the table.

In his first chapter of Part I ("This World"), Abbott (2002: 34–35) provides a description and an accompanying graphic to describe Flatland. It is instructive to reproduce it here:

> I call our world Flatland, not because we call it so, but to make its nature clearer to you, my happy readers, who are privileged to live in Space.
>
> Imagine a vast sheet of paper on which straight Lines, Triangles, Squares, Pentagons, Hexagons, and other figures, instead of remaining fixed in their places, move freely about, on

or in the surface, but without the power of rising above or sinking below it, very much like shadows—only hard and with luminous edges—and you will then have a pretty correct notion of my country and countrymen. Alas, a few years ago, I should have said "my universe": but now my mind has been opened to higher views of things.

In such a country, you will perceive at once that it is impossible that there should be anything of what you call a "solid" kind; but I dare say you will suppose that we could at least distinguish by sight the Triangles, Squares, and other figures, moving about as I have described them. On the contrary, we could see nothing of the kind, not at least so as to distinguish one figure from another. Nothing was visible, nor could be visible, to us, except Straight Lines; and the necessity of this I will speedily demonstrate.

Place a penny on the middle of one of your tables in Space; and leaning over it, look down upon it. It will appear as a circle.

But now, drawing back to the edge of the table, gradually lower your eye (thus bringing yourself more and more into the condition of the inhabitants of Flatland), and you will find the penny becoming more and more oval to your view; and at last when you have placed your eye exactly on the edge of the table (so that you are, as it were, actually a Flatlander) the penny will then have ceased to appear oval at all, and will have become, so far as you can see, a straight line.

The same thing would happen if you were to treat in the same way a Triangle, or Square, or any other figure cut out of pasteboard. As soon as you look at it with your eye on the edge on the table, you will find that it ceases to appear to you a figure, and that it becomes in appearance a straight line. Take for example an equilateral Triangle—who represents with us a Tradesman of the respectable class. Figure 1 represents the Tradesman as you would see him while you were bending over him from above; Figures 2 and 3 represent the Tradesman, as you would see him if your eye were close to the level, or all but on the level of the table; and if your eye were quite on the level of the table (and that is how we see him in Flatland) you would see nothing but a straight line.

The whole novel is, in a sense, an experiment in spatial reasoning, simultaneously raising intriguing philosophical questions about space and how the human brain perceives and interprets spatial information. For example: Is there a *formal* ("form-based") relation between our three-dimensional world—called Sphereland—and the two-dimensional one? The answer is yes, because a Sphereland figure can be changed into a Flatland one, and vice versa, by performing a specific kind of alteration to it. Consider a three-dimensional cube made up of six sides, which are, geometrically speaking, six equal Flatland squares. The cube can be easily transformed into a Flatland figure by simply "unfolding" it as shown in Fig. 4.6.

The resulting Flatland figure can, of course, be just as easily transformed back into a three-dimensional cube by folding the six squares together. So, another way to envision this transformation is in terms of assembly. Now, the question that such folding and unfolding raises is truly a mind-boggling one: Can an analogous transformational method be envisioned that would produce a four-dimensional cube? The answer produces a *hypercube*, which consists of eight cubes joined together, as shown in Fig. 4.7.

The phases that produce a hypercube are illustrated in Fig. 4.8.

Fig. 4.6 Cube and its unfolded version

Cube (three-dimensional folded version)

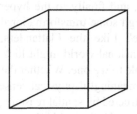

Cube (two-dimensional unfolded version)

Fig. 4.7 Hypercube (Wikimedia Commons)

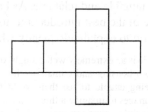

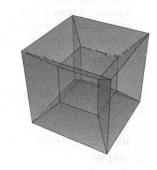

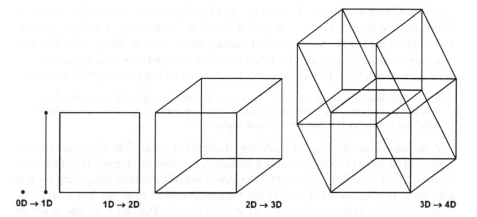

0D → 1D   1D → 2D   2D → 3D   3D → 4D

Fig. 4.8 Hypercube construction (Wikimedia Commons)

The construction technique shows how we go from the line to the square, to the cube, and finally to the hypercube, indicating that the dimensions are connected through some transformational principle. Is the hypercube truly a four-dimensional figure? Like the Flatlanders, we Spherelanders can only infer what a four-dimensional world might look like by reasoning transformationally. We will never be able to *see* one. Whether the hypercube really is a four-dimensional form or not in any meaningful physical sense is actually beside the point. The fact that it can be constructed by spatial reasoning is truly mind-boggling. While our viewing perspective may be limited by living in a three-dimensional space, our imagination is not.

Flatland presents advanced geometrical ideas in narrative form, thus making them tangible and relatable. As Freiberger (2006) has aptly observed, the novel "still is one of the best introductions to a mathematical world of higher dimensions." She goes on to explain eloquently why:

> This is an extremely well-thought-out story; every aspect of life in Flatland is accounted for, from housing and climate to the way in which Flatlanders recognise each other's shape (being unable to see their world from above). Abbott's descriptions of how the square manages to imagine a three-dimensional world are a great guide to how we might go about imagining four (or more) dimensions. In fact, the square even conjectures that a four-dimensional world might exist, much to the annoyance of the sphere which considers itself supreme. Abbott's analogy is clear and strong, and will make sense to the most unmathematically minded reader. It's a beautiful defense of mathematical thought and its power to open doors to fascinating new worlds.

As Hofstadter (1979, Hofstadter and Sander 2013) has argued cogently, imaginative analogies are the key to understanding how math ideas come into being or at least coagulate in the mind. Analogies emanate from an interaction between abduction and reasoning, as discussed, and which has been called the bi-part mind. When these become part of a cognitive flow that leads to abstract models, the brain is in a position to discover new ways of perceiving (literally) the world around and beyond us.

However, there is always a caveat in such thinking, as we have seen in the case of some puzzles and paradoxes. As discussed in the previous chapter, one of the main premises of formal analysis is that any set of mathematical propositions must be complete (leaving out contrasting possibilities) and consistent (avoiding circularities, ambiguities, and statements that cannot be proved or disproved). Euclid's geometry was the first approach toward ensuring completeness and consistency in mathematical thinking. However, his fifth axiom turned out to be a problematic one:

> If a straight line crossing two straight lines makes the interior angles on the same side less than two right angles, the two straight lines, if extended indefinitely, meet on that side on which are the angles less than the two right angles.

Also known as the Parallel Postulate, it attracted immediate criticism, since it seemed to be more of a theorem than a postulate. Proclus (in Morrow 1970) wrote as follows: "This postulate ought even to be struck out of Postulates altogether; for it is a theorem. It is impossible to derive the Parallel Postulate from the first four."

In the 1800s, mathematicians finally proved that the Parallel Postulate is not an axiom. This led to the creation of geometric systems in which Euclid's postulate was revised. From this, non-Euclidean geometries emerged. Like Abbot's novel these literally imagined geometry in diverse new ways. By applying the triangle form to a globe, the angles in the triangle will be greater than $180°$, which contravenes a fundamental principle of Euclidean geometry. In Lobachevskian geometry, moreover, the Parallel Postulate is replaced by the following one: *Through a point not on a given line, more than one line may be drawn parallel to the given line.* In one model of n-dimensional geometry, the plane is defined as a set of points that lie in the interior of a circle. Parallel lines are defined, of course, as lines that never intersect. Around 1860, Riemann had another whimsical hunch: Is there a world where no lines are parallel? The answer is the surface of a sphere on which all straight lines emanating from the poles are great circles. It is, in fact, impossible to draw any pair of parallel lines on the surface, since they would meet at the two poles.

Because one important use of geometry is to describe the physical world, we might ask which type of geometry, Euclidean or non-Euclidean, provides the best model of reality. Some situations are better described in non-Euclidean terms, such as aspects of the theory of relativity. Other situations, such as those related to building, engineering, and surveying, seem better described by Euclidean geometry. In other words, Euclidean geometry is still around because it is relevant to Flatland and many sectors of Sphereland. Lobachevskian and Riemannian geometries, and by extension n-dimensional geometries, have applications to "unseen" domains of reality. Even so, the latter are analogical extensions of Euclidean ideas, as Lewis Carroll (1879) persuasively argued. Abbott's novel ingeniously showed how analogy is the crux to understanding and extending geometry.

Loyd's Optical Illusion Puzzles

One of the most ingenious Gotcha puzzles based on the assembly-movement archetype, which is intended to dupe spatial reasoning itself, is Sam Loyd's Get Off the Earth Puzzle. The puzzle is a "cut-and-slide" optical illusion. The idea underlying its construction probably goes back to a puzzle included in the book titled *Rational Recreations* by William Hooper (1782). Loyd created his version by fastening a smaller paper circle to a larger one with a pin so that it could spin around. Then, with appropriate artwork on both circles, he made the figure look like the Earth with thirteen "Chinese warriors" on it. Loyd patented his puzzle in 1897. It sold more than ten million units in that year alone:

> When the smaller circle is turned slightly, as shown below, the thirteen warriors turn mysteriously into twelve. Where did the thirteenth warrior go?

With the world oriented so that the large arrow on it points to the N.E. point on the background, 13 men can be counted. But when the Earth is turned slightly, so that the arrow points to the N.W. point, there are only 12 such men. See Fig. 4.9.

When the Earth is rotated, the pieces are rearranged in such a way that each of the Chinese Warriors gains a sliver from his neighbor. For example, at the lower left

Fig. 4.9 Loyd's Get off the Earth Puzzle (1897)

Fig. 4.10 Loyd's rectangle
optical illusion (1914)

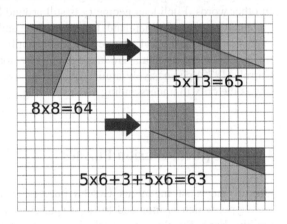

on the left diagram, there are two Warriors next to each other. The top one is missing
a foot. When the Earth is rotated, he gains a foot from his neighbor on the right. That
neighbor gains 2 ft (since he lost one) and one small piece of a leg. And so on and so
forth. As a result of the rotation, one of the Warriors will "lose" all his parts, making
it seem that he has "disappeared."

Loyd presented another well-known optical illusion puzzle of this kind, again
traceable to Hooper's work and then discussed mathematically in the 1868 issue of
the *Zeitschrift für Matematik* (see Petkovic 2009: 14). Loyd included it in his
Cyclopedia of Tricks and Puzzles of 1914. He divided an 8×8 square with area
of 64, outlining two triangles and two trapezoids in it (as shown in Fig. 4.10). When
he reassembled the pieces, as shown, he got a 5×13 rectangle with an area of
65, which is one unit more, and when he rearranged the same pieces as shown in the
last diagram he got an area of 63, which is one unit less.

How did the extra small square get in there? And how did it disappear? The truth is that the edges of the figures as assembled do not actually coincide along either diagonal. A magnification would show that the diagonal is really a long and very narrow parallelogram that can barely be noticed. But this is not the end of the matter. If we subtract the area of the above rectangle—$5 \times 13 = 65$— from the area of the original square $8^2 = 64$, we get, of course, the difference of "1"—which represents the area of the missing square. Let's write this out as follows:

Area of rectangle:	5×13
Area of original square:	8^2
Difference between the two areas:	$(5 \times 13) - 8^2 = 1$

Now, if we look closely at the actual digits in the last expression—5, 8, and 13— we can see that they are three consecutive numbers in the Fibonacci sequence. If we dissect squares of dimensions $3^2, 21^2, 55^2$ in the same way that we dissected our 8^2 figure, we will produce rectangles of dimensions 5×2, 13×3, and 34×89 by rearrangement. All the digits in these expressions belong to the Fibonacci sequence. In each case an extra little square unit is produced in the rearrangement process. Remarkably, subtracting the rectangles from the original squares in the same way as we did above also produces consecutive Fibonacci numbers:

$(5 \times 2) - 3^2 = 1$ $\rightarrow \ldots 2, 3, 5 \ldots$
$(13 \times 34) - 21^2 = 1$ $\rightarrow \ldots 13, 21, 34, \ldots$
$(34 \times 89) - 55^2 = 1$ $\rightarrow \ldots 34, 55, 89 \ldots$

These serendipitous interconnections among puzzles can only occur, as discussed throughout, if the same type of archetypal structure is involved.

The Four-Color Problem

A puzzle that has had enormous implications for mathematics arose initially from the observations of mapmakers. It is known as the Four-Color Problem. Mapmakers had believed from antiquity that four colors were sufficient to color any map, so that no two contiguous regions would share a color. This insight caught the attention of mathematicians in the nineteenth century, after a young student at University College, London, named Francis Guthrie suspected in 1852 that four colors would always be enough, although there is some evidence that Augustus Möbius had already discussed this hypothesis in a lecture to his students as far back as 1840. Guthrie apparently wrote about the problem to his younger brother, Frederick. The story then goes that Frederick described it to his own professor, the prominent British mathematician and puzzlist, Augustus De Morgan, who quickly realized that the Four-Color Problem had many important ramifications for mathematics. Word of the problem spread quickly.

The Four-Color Problem is not a puzzle in the traditional sense, since it arose initially as a practical problem within cartography. Nevertheless, it has all the structural features of a puzzle, especially since the answer seemed iitially to be

Fig. 4.11 Diagram
showing that four colors will
suffice

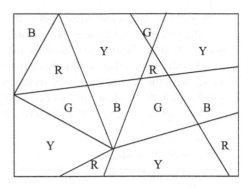

elusive. It requires a large dose of spatial reasoning not only to solve, but just to envision. In its simplest form it reads as follows:

> What is the minimum number of tints needed to color the regions of any map distinctively?
> (If two regions touch at a single point, the point is not considered a common border).

Of course, a two-region diagram will require two colors and a three-region one, three. No map consisting of more regions has been found that requires more than four colors. Figure 4.11 shows blue (B), red (R), green (G), and yellow (Y) colors that are sufficient to color all its regions distinctively.

The challenge is to prove that this is always the case, no matter what diagram is used—that is, the challenge is to prove that four colors are *sufficient* to color any map, no matter how many regions it has. After De Morgan made the Four-Color Theorem widely known, mathematicians started in earnest trying to prove it with the traditional methods of proof. But their efforts were consistently fruitless. A proof finally came from two American mathematicians, Wolfgang Haken and Kenneth Appel. It was seen at first to be peculiar because it broke with tradition; it used an algorithm, which essentially checked to see if a map could be colored by more than four tints (Haken 1977, Haken and Appel 1977, 2002). Haken and Appel themselves admitted that their proof may not be the last word on the problem (2002: 193): "One can never rule out the chance that a short proof of the Four-Color Theorem might some day be found, perhaps by the proverbial bright high-school student."

The Haken-Appel proof can be summarized as follows (paraphrased from Wilson 2002):

1. They demonstrated that no single map in a set of 1936 maps constituted a counterexample to the Four-Color Theorem, using an algorithm designed to look for counterexamples.
2. Any map that could potentially constitute a counterexample must show a portion that looks like any one of the 1936 maps. No one was found.
3. Showing (2) involved hundreds of pages of hand (non-computer) analysis; in other words, the proof was based on "brute force" analysis.
4. Haken and Appel concluded that no counterexample exists because any such case must contain, and at the same time not contain, one of the 1936 maps—which is a contradiction.
5. The contradiction means that there are no counterexamples—hence the proof.

The program Haken and Appel wrote was thousands of lines long and took 1200 hours to run. Since then, mathematicians have checked the program, finding only minor and fixable problems in it. The two mathematicians essentially showed that no exception to the Four-Color Theorem will ever be found. However, to this day, some mathematicians remain uncomfortable with their demonstration, for the simple reason that it is based on an algoritthm, rather than on a traditional proof. Soon after the publication of the Haken-Appel algorithm, Thomas Tymoczko (1979) encapsulated the feelings of some mathematicians when he observed that, if accepted, Haken and Appel's work puts mathematics in a position to radically alter its traditional view of proof. By accepting a new way of proving the theorem, mathematics did indeed undergo a paradigm shift, as Tymoczko suggests. The point is that this was due essentially to a seemingly simple spatial reasoning puzzle derived from observing the coloring of maps.

As Charles Peirce, who came under its spell, so aptly put it in a lecture he delivered at Harvard in the 1860s, the problem is so infuriating precisely because it appears to be so simple to prove, and yet, to this day, no one has found a proof for it with the traditional methods . Whether conducted by human reasoning alone or with the assistance of a computer, proof is at the core of mathematics. For a mathematical statement or theorem to be accepted as true it must be proved, otherwise it remains a conjecture. In a new radical way, the Haken-Appel proof established that the notion of proof in mathematics is hardly static, but is constantly evolving.

The more elusive a proof is, the more it is hunted down, even if it may seem to have no implications above and beyond the satisfaction of finding a proof in itself. To use one recent example from topology, consider the conjecture identified by Henri Poincaré in 1904—if any loop in a certain kind of three-dimensional space can be shrunk to a point without ripping or tearing either the loop or the space, then the space is equivalent to a sphere. To put it another way, Poincaré suggested that any topological form without holes in it has to be a sphere. To grasp the conjecture, one could imagine stretching a rubber band around a ball. The band can be contracted slowly, so that it neither breaks nor loses contact with the ball. The band cannot be shrunk to a point if it is stretched around a doughnut, whether around its hole or body. It can be done, however, with any topological equivalent of a ball, such as a deformed melon, a baseball bat with bulges, and the like. The surface of the ball, but not of the doughnut, is "simply connected." Any simply connected two-dimensional closed surface, however distorted, is topologically equivalent to the surface of a ball. Poincaré wondered if simple connectivity characterized three-dimensional spheres as well.

His conjecture was finally proved by Russian mathematician Grigory Perelman in 2002, posting his solution on the Internet (O'Shea 2007, Gessen 2009). It is much too complex to discuss here (being over 400 pages). Suffice it to say that an analysis of the proof shows that it involves analogies, connections, hunches, and various types of proof, forming an interweaving array of imaginative and rational thinking. As Chaitin (2006: 24) puts it, "mathematical facts are not isolated, they are woven into a spider's web of interconnections." And as Wells (2012: 140) states:

Proofs do far more than logically certify that what you suspect, or conjecture, is actually the case. Proofs need ideas, ideas depend on imagination and imagination needs intuition, so proofs beyond the trivial and routine force you to explore the mathematical world more deeply—and it is what you discover on your exploration that gives proof a far greater value than merely confirming a fact.

The Assembly Archetype

An episode that parallels but predates Poincaré's enigma is Kepler's conjecture (Szipro 2003). In 1611, Kepler suggested that the optimal way to pack spheres as densely as possible was to pile them up in the same way that grocers stack oranges or tomatoes in crates. Proving it was to be as difficult as any of the great problems of mathematics that seem to have a simple, but elusive, proof. Thomas Hales and Samuel Ferguson have put forth what is now a definitive proof (Hales 1994, 2000, 2005, Hales and Ferguson 2006, 2011). At first, the conjecture seemed to emanate from beyond the realm of possibility. But the Hales-Ferguson proof, like the Perelman one, was constructed as an amalgam of imaginative inferences, practical ideas, and various types of proof. Initially, Kepler's conjecture aimed to show that no assembly of equally sized spheres filling three-dimensional space has a greater average density than that of the cubic close packing (face-centered cubic) and hexagonal arrangements.

The Hales-Ferguson proof is proof by exhaustion, which involves checking individual cases using a computer algorithm, much like the Haken-Appel proof. Kepler actually turned the packing problem into a spatial one: What assembled figure gives us optimization density in regard to the packed spheres? The answer, as the Hales-Ferguson proof demonstrates, comes through a blend of imaginative insights with trial and error.

Like the recreational games discussed in this and previous chapters, the Kepler conjecture is based on an assembly archetype, which can be explicated simply as the mind's desire to assemble scattered pieces into meaningful wholes. Packing, tiling, and assembly puzzles are all models of such archetypal thinking. In 1957, for instance, Solomon W. Golomb invented the game of *polyominoes*. A polyomino is a two-dimensional shape formed by joining square, flat pieces of paper or plastic. There are several basic types of polyominoes: the *domino* is a two-square figure, the *tromino* a three-square figure, the *tetromino*, a four-square figure, and the *pentomino* a five-square figure. They can be assembled into a variety of forms. For instance, below are pentominoes that have been assembled to make various figures (Fig. 4.12):

Fig. 4.12 Pentominoes
(Wikimedia Commons)

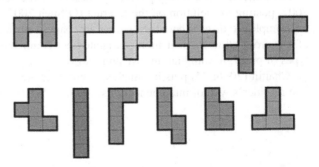

Fig. 4.13 Rectangular
pentominoes (Wikimedia
Commons)

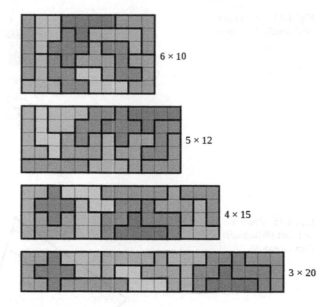

The degree of spatial-imaginative thinking that is required to solve
derived polyomino puzzles is truly high. Figure 4.13 contains a few examples of
how the pieces can be assembled to make rectangular figures

Mathematically, it has been found that there is a limit to the number of
polyominoes that can be assembled meaningfully with the following boundaries
(n = number of polyominoes): $3.72^n < P(n) < 4.65^n$ (Klarner 1967). In turn, this type
of analysis has had implications for estimating cell growth— an unexpected serendipitous outcome of this puzzle archetype.

In 1936, the Danish poet-mathematician Piet Hein had already invented a three-
dimensional version of Golomb's two-dimensional game, which he called *Soma
Cubes*. The game was introduced to recreational mathematicians in Gardner's
Scientific American column in 1958 (Gardner 1998: 70–71). As with polyominoes,
the goal of Hein's game is to join the cubes together to form shapes. In Fig. 4.14. are
examples of assemblages that can be made with Soma cubes.

It has been found that there are 240 possible distinct shapes that can be
formed with the cubes. But there is only one way that the "T" piece can be
assembled. The proof resembles the same kind of reasoning used by Euler with
regard to his Königsberg Bridges Puzzle. Essentially, a T-shape must fill two
corners, and there is only one orientation (without rotations and reflections) in
which it does that.

Hein is also known for the game of Hex in 1942, which was independently
invented by the great mathematician John Nash—a serendipitous fact that, as
discussed throughout this book, suggests archetypal thinking. This is played on a
diamond-shaped board made up of interlocking hexagons (hence the abbreviated
name Hex). A typical board has eleven hexagons on each edge. One player has a
supply of black counters, the other an equal number of white. The players take turns

Fig. 4.14 Soma cubes
(Wikimedia Commons)

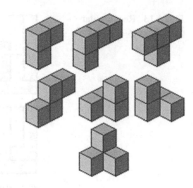

Fig. 4.15 Penrose's kite
and dart (Wikimedia
Commons)

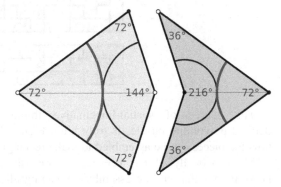

putting one of their counters on any unoccupied hexagon. The object is to complete a continuous chain of counters from one edge of the board to the opposite edge. Players try to block each other's attempts to construct a chain, as well as trying to complete their own. Recalling Huygen's observation from Chap. 3 that some games might harbor within them some "deep theory," Hex is serendipitously connected to Brouwer's fixed-point theorem. This was shown by David Gale (1979), who proved that no game can end in a draw, which is equivalent to the Brouwer fixed-point theorem for the plane. By considering n-dimensional versions of Hex, one can prove in general that Brouwer's theorem is equivalent to the theorem for playing Hex.

In 1974, the British physicist Roger Penrose invented a different kind of assembly game, which has come to be called, appropriately, the Penrose Tile Game. It involves making nonperiodic tilings of the plane with two figures. A periodic tiling is a design that recurs horizontally or vertically across the plane. The grid design of a sheet of graph paper is an example of such a tiling, since the same pattern of white square after white square recurs both vertically and horizontally. A nonperiodic tiling does not recur either vertically or horizontally. The Penrose Tile Game produces an infinite number of different nonperiodic tilings of a surface, with two figures called a *dart* and a *kite*. Incidentally, Penrose used the golden ratio to design his darts and kites. See Fig. 4.15.

As many kites and darts as desired can be made. The goal of the game is to join them to produce different kinds of figures, such as suns and stars. In a basic sense,

Fig. 4.16 A Penrose tiling
(Wikimedia Commons)

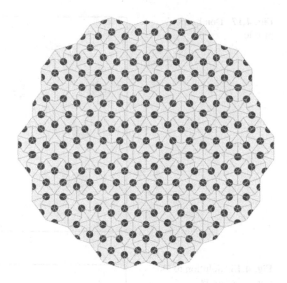

Penrose's tiles, like Golomb's polyominoes or Hein's Soma Cubes, are works of "pure geometrical art," inviting the eye to detect pattern and symmetry for their own sake. They produce the same kind of aesthetic pleasure that comes from looking at a cubist-type painting. As an example, Fig. 4.16 contains a Penrose tiling.

Spatial Reasoning

Since antiquity, constructing and dissecting figures with two instruments—the ruler (or straightedge) and the compass—have always constituted concrete techniques for investigating the properties of geometrical forms and for deducing theorems about them. As we saw above, Loyd's Get Off the Earth Puzzle conceals within it a simple construction technique, which never fails to stupefy people who are unaware of how Loyd created his illusion. This is perhaps why the puzzle typically produces the Gotcha effect. Together with the other puzzles described here, it shows how our imagination not only helps us discover deeply embedded archetypes, such as the assembly one, but also how it plays mischievously on our perception.

As a final example, consider the following puzzle devised by Don Lemon (now thought to be a pseudonym) in his 1890 *Everybody's Book of Illustrated Puzzles* (adapted from Costello 1988: 11). See Fig. 4.17:

> Five boarders live in a house with a garden. The owner of the house wants to divide the garden among the five. There are ten trees in the garden, laid out in a particular way. How can the owner divide it so that each of the five boarders receives an equal share of the garden and two trees?

Fig. 4.17 Don Lemon's
puzzle

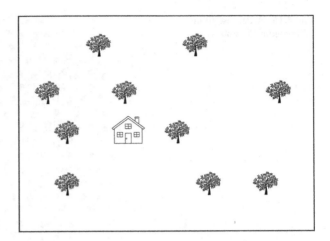

Fig. 4.18 Solution to Don
Lemon's puzzle

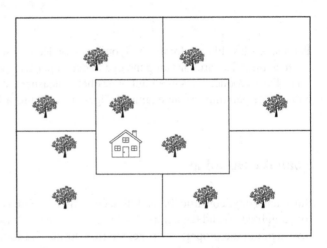

Coming up with a solution requires an Aha insight, bolstered by imaginative spatial reasoning, and is thus reflective of how any assembly puzzle is solved, namely by envisioning the parts of a situation and how these can be combined (or partitioned) in some specific fashion. After some trial and error and some visual reasoning the solution is shown in Fig. 4.18.

Puzzles like this one are "miniature blueprints" of how Aha thinking is wired into the brain and of how it allows us to come up spontaneously with solutions to seemingly intractable conundrums. Based on a general assembly archetype, they impel us to find an order among the pieces by arranging them into holistic figures.

The Pythagoreans were petrified by anomalies in spatial configurations. They considered their "theory of order" seriously undermined when, ironically, Pythagoras's own theorem revealed the existence of irrational numbers such as $\sqrt{2}$. This "chaotic" number stared them straight in the face each time they drew an isosceles right-angled triangle with equal sides of unit length. The length of its hypotenuse was the square root of the sum of $1^2 + 1^2$, or $\sqrt{2}$, a number that cannot be represented as the ratio of two integers, or as a finite or repeating decimal. For the Pythagoreans, rational numbers had a "rightness" about them; irrational ones such as $\sqrt{2}$ did not. And yet, there they were, defying logic and sense, and challenging the system of order that the Pythagoreans so strongly desired to establish. So disturbed were they that, as some stories would have it, they "suppressed their knowledge of the irrationality of $\sqrt{2}$, and went to the length of killing one of their own colleagues for having committed the sin of letting the nasty information reach an outsider" (Ogilvy 1956: 15). The colleague is suspected to have been Hipassus of Metapontum.

Incredibly, even in the domain of chaos, the human mind finds ways of making sense of things. Assembly puzzles and games are miniature mirrors of how we search for pattern and its many serendipitous manifestations. We require, existentially, that there must be order in the scheme of things. *Chaos theory*, founded by Henri Poincaré, has shown, in fact, that there are patterns even in random events. Already in the 1960s, simplified computer models demonstrated that there was a hidden structure in the seemingly chaotic patterns of the weather. When these were plotted in three dimensions they revealed a butterfly-shaped fractal set of points. Similarly, leaves, coastlines, mounds, and other seemingly random forms produced by Nature reveal hidden fractal patterns when examined closely.

All this constitutes a truly profound existential paradox. Why is there order in chaos? And more importantly: Why do we constantly seek order and pattern, as the assembly puzzles in this chapter imply? Perhaps the ancient myths provide, after all, the only plausible response to this question. According to the *Theogony* of the Greek poet Hesiod (eighth century BCE), Chaos was produced by Earth, from which arose the starry, cloud-filled Heaven and its basis in Order. In a later myth, Chaos was portrayed as the formless matter from which the Cosmos, or harmonious Order, was created. In both versions, the ancients felt deeply that order arose out of chaos. This feeling continues to reverberate today, and is perhaps behind the sense of satisfaction we get when solving a geometrical puzzle or finding our way through a maze. For some truly mysterious reason, our mind requires that there be order within apparent disorder. This is, in fact, a deeply-embedded archetype that mainifests itself in all sorts of assembly puzzles, as surmised in this chapter.

Chapter 5
The Mathematical Mind

*I know that two and two make four and should be glad to
prove it too if I could—though I must say if by any sort of
process I could convert 2 & 2 into five it would give me much
greater pleasure.*

—Lord Byron (1788–1824)

As argued throughout this book, puzzles have played as much a role as any other
human artifact, mental tool, or device in human history as sparks for discovery. The
Ahmes Papyrus is more than a source of ancient mathematics. It is the first text to
show, rather conspicuously, that puzzles and mathematics have a common origin. As
mentioned, each puzzle in the work is both a creative (dialectic) conundrum and a
mini-treatise in mathematical thinking. This has been called "Ahmes' legacy" in this
book, which claims above all else that puzzles are mirrors of the inner workings of
the mathematical mind. The classic puzzles of Ahmes, Alcuin, Fibonacci, Euler,
Cardano, Lucas, Carroll, and many others are miniature models of that mind,
showing how the flow of thought goes from experience, to imaginative hunches,
and then on to a solution. Once the mathematical archetype is extracted from the
solution via generalization, the puzzle becomes a kind of intellectual meme that
makes its way into other mathematical minds to suggest new ways of doing
mathematics. Mathematical cognition can thus be characterized as a blended form
of imaginative-reflective thinking (Poe's bi-part soul) that is sparked by the imagination's interpretation of experiences via their serendipitous connectivity. This view
is consistent with the neuroscientific work being conducted on so-called blending
theory today (Fauconnier and Turner 2002, Danesi 2016), whereby the brain is seen
as an organ that connects imaginative thoughts with each other in order to produce a
new unit of thought. In effect, the meaning of something is not in its individual parts,
but in the way they are connected or combined.

Starting specifically with Bachet, certain puzzles have been devised and used by
mathematicians to exemplify or experiment with mathematical ideas, by connecting
(blending) them in ingenious ways, simultaneously establishing recreational mathematics as both a branch of theoretical mathematics and of cognitive science. Take,

© Springer International Publishing AG, part of Springer Nature 2018
M. Danesi, *Ahmes' Legacy*, Mathematics in Mind,
https://doi.org/10.1007/978-3-319-93254-5_5

as another example, the following well-known simple puzzle, which describes a physical situation that can potentially arise in real life:

> I have six billiard balls, one of which weighs less than the other five. Otherwise they all look identically the same. How can I identify the one that weighs less on a balance scale with only two weighings?

Weighing puzzles such as this one are best approached with trial runs using fewer items, thus setting the experiential basis for solving them. This initial phase will allow us to see if there is some general principle or procedure that can be applied to solving the original puzzle and for extending it to more complex versions. For this particular puzzle, it is advisable to consider first the weighing of two balls. Needless to say this is a trivial test, since both balls can be put on the pans of the scale at the same time—one on the left pan and one on the right pan. But it still highlights what will happen. The pan that goes up, of course, is the one holding the ball that weighs less. In this case, one weighing was enough to identify the culprit ball.

Next, we consider the weighing of four balls. First, we divide the four balls equally in half: that is, into two sets of two balls each.

WEIGHING 1
   We put two balls on the left pan and two on the right pan at the same time. The pan that goes up contains the ball that weighs less, but we do not yet know which one of the two.
   WEIGHING 2
   So, we take the two suspect balls from the pan that went up, discarding the ones on the other pan. We put each one of the suspect balls on a separate pan—one on the left pan and one on the right pan. The pan that goes up contains the ball that weighs less.

These two trial runs have shown us how to go about methodically identifying the suspect ball in a collection of an even number of balls. We are now ready to turn our attention to the original puzzle. There are six balls in the collection, and we are told to identify the suspect ball in only two weighings. We start off in the same way as we did before: that is, we divide the six balls equally in half, as two sets consisting of three balls each. Then, we go ahead and perform the first weighing as before:

WEIGHING 1
   We put three balls on each pan this time—three on the left pan and three on the right pan. The pan that goes up contains the ball that weighs less, but we do not yet know which one of the three.

Now, for the second weighing, we focus our attention on the set containing the suspect ball, discarding the ones on the other pan. Is it possible to identify the suspect ball in just one more weighing (recall that the puzzle asks us to identify it in just two weighings)?

WEIGHING 2
   We select any two of the three balls to weigh, putting the third ball aside. We put each one on a separate pan—one on the left pan and one on the right pan. What are the possible outcomes of this second weighing? If they balance then the suspect ball is the one put aside; if one of the pans goes up, then it contains the suspect ball. These are the only outcomes; so, one way or the other, we have identified the suspect ball.

It took just two weighings, and a little bit of clever Aha thinking, to identify the ball that weighs less. The interesting thing to note is that we penetrated the structure of the puzzle by trying out different versions. The next thing to do is to examine what it entails mathematically. In other words can we generalize the structure of the puzzle? We simply continue the "weighing experiment" with more and more balls in different numerical collections in order to flesh out some general pattern and then to test this against other similar patterns in other areas of mathematics via inductive proof. That is how many parts of recreational mathematics unfold. Called the Generalization Principle here, it characterizes the blend of imaginative and reflective thinking that work in tandem to produce ideas, concepts, and perhaps even branches (as we have seen).

In this final chapter, the focus on the relation between puzzles and the mathematical mind will be discussed synthetically. A few loose thematic strings will thus be tied together by considering a few other classic puzzles—the Traveling Salesman Problem, the Monty Hall Problem, and logic games such as Sudoku.

The Traveling Salesman Problem

The Traveling Salesman Problem (TSP) is an important one for several areas of mathematics. It is based on the same kind of archetype that characterizes Euler's Königsberg Bridges Puzzle and other graph network puzzles. It has a specific kind of structure that can be easily modeled on a computer. Here's a standard version of the problem by Benjamin, Chartrand, and Zhang (2015: 122).

> A salesman wishes to make a round-trip that visits a certain number of cities. He knows the distance between all pairs of cities. If he is to visit each city exactly once, then what is the minimum total distance of such a round trip?

In this puzzle, all the vertices of a Eulerian graph are to be used. The crux of the solution to the TSP is elaborated comprehensively by Benjamin, Chartrand, and Zhang (2015: 122) as follows (where $c = a$ city, $n =$ number of vertices in a graph). Given its clarity, it is worth reproducing here:

> The Traveling Salesman Problem can be modeled by a weighted graph G whose vertices are the cities and where two vertices u and v are joined by an edge having weight r if the distance between u and v is known and this distance is r. The *weight of a cycle C* in G is the sum of the weights of the edges of C. To solve this Traveling Salesman Problem, we need to determine the minimum weight of a Hamiltonian cycle in G. Certainly G must contain a Hamiltonian cycle for this problem to have a solution. However, if G is complete (that is, if we know the distance between every pair of cities), then there are many Hamiltonian cycles in G if its order n is large. Since every city must lie on every Hamiltonian cycle of G, we can think of a Hamiltonian cycle starting (and ending) at a city c. It turns out that the remaining $(n - 1)$ cities can follow c on the cycle in any of its $(n - 1)!$ orders. Indeed, if we have one of the $(n - 1)!$ orderings of these $(n - 1)$ cities, then we need to add distances between consecutive cities in the sequence, as well as the distance between c and the last city in the sequence. We then need to compute the minimum of these $(n - 1)!$ sums. Actually, we need *only* find the minimum of $(n - 1)!/2$ sums since we would get the same sum if a sequence was traversed in reverse order. Unfortunately, $(n - 1)!/2$ grows very, very fast. For example, when $n = 10$, then $(n - 1)!/2 = 181,400$.

Fig. 5.1 A solution to the
TSP (Wikimedia Commons)

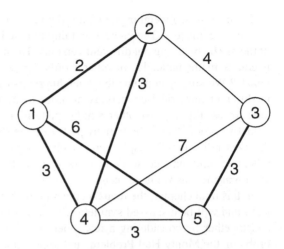

By translating the physical referents of the problem (distances, cities, and so on) into mathematically symbolic notions, such as paths, weights, and so on that apply to graphical network systems, a mathematical model of the TSP can be set up—a model that decomposes all aspects of the problem into its essential parts. Because of this, it can be translated into an algorithm based on a Hamiltonian cycle. Figure 5.1 shows one solution that connects every dot to produce the shortest route.

For historical accuracy, it should be mentioned that the TSP was first presented in the 1930s as an early challenging problem in algorithmic optimization. As Bruno, Genovese, and Improta (2013: 201) note:

> The first formulation of the TSP was delivered by the Austrian mathematician Karl Menger who around 1930 worked at Vienna and Harvard. Menger originally named the problem the messenger problem and set out the difficulties as follows. At this time, computational complexity theory had not yet been developed: "We designate the Messenger Problem (since this problem is encountered by every postal messenger, as well as by many travelers) the task of finding, for a finite number of points whose pairwise distances are known, the shortest path connecting the points. This problem is naturally always solvable by making a finite number of trials. Rules are not known which would reduce the number of trials below the number of permutations of the given points. The rule, that one should first go from the starting point to the point nearest this, etc., does not in general result in the shortest path."

Of course, Menger's challenge has been tackled rather successfully by computer science and mathematics working in tandem today with the development of the field of combinatorial optimization, which was developed to solve problems such as the TSP one. In 1954 an integer programming formulation was developed to solve the problem alongside the so-called "cutting-plane" method, "which enables the finding of an optimal solution (namely, the shortest Hamiltonian tour) for a TSP involving 49 U.S. state capitals" (Bruno, Genovese, and Improta 2013: 202). The problem has been generalized in various ways and studied algorithmically, leading to the growth of optimization theory.

Interestingly, it has had several serendipitous applications, such as in the area of DNA sequencing. For the present purposes it is sufficient to say that it shows how to attack NP-hard problems in general (Bruno, Genovese, and Improta 2013: 205–207). The tactics in the attack include the following two:

1. Creating algorithms for finding solutions.
2. Devising heuristic algorithms that may not provide a solution but will shed light on the problem and generate interesting subproblems in the process.

Algorithms proceed one step at a time, from a starting point to an end-point. They are based on giving a logical form to a problem's structure. This is why the algorithm mirrors the sequential organization of a traditional proof, in that the moves from one step to the other are computable. Computability is thus a metric of solvability and provability.

Needless to say, the TSP is connected to previous puzzles that involve graphs, networks, and arrangements possibilities, from the River-Crossing Puzzles and the Königsberg Bridges Puzzle to many others based on the same archetype. In a phrase, the original insights in a puzzle have psychologically "catastrophic" consequences, to use René Thom's concept discussed earlier. When looked at cumulatively, this suggests that the mathematical mind is a highly imaginative one that generates ideas in the form of puzzles, problems, and games, extracted from observation and experience, that become more and more complex as they are connected with other domains of mathematics.

The TSP is a problem in Hamiltonian analysis, which is itself a descendant of Eulerian graph theory. The TSP is decidable, and is often enlisted to show what this entails. As Elwes (2014: 289) puts it: "If P $\neq$ NP, then there is some problem in NP which cannot be computed in polynomial time. Being NP-complete, the Travelling Salesman Problem must be at least as difficult as this problem, and so cannot lie in P" (see Cook 2014).

The *Hamiltonian Cycle* is named after the Irish mathematician William Rowan Hamilton, who presented this concept as a game that he called *Icosian*. The object is to find a cycle along the edges of a dodecahedron such that every vertex is reached a single time, and the ending point is the same as the starting point. The puzzle-game was distributed commercially as a pegboard with holes at the nodes of the dodeca-hedral graph and was subsequently marketed in Europe in many versions. The solution is a cycle containing twenty (in ancient Greek *icosa*) edges. Figure 5.2 illustrates a simple Icosian grid.

Hamiltonian cycles must be examined individually, and finding a Eulerian path in them—if one exists—is a matter of trial and error. The surface is a closed one with no break. If we start a path anywhere that cuts the surface at any point, we will end up at the point from which we started. No matter where we penetrate the surface, we are still outside of it. In the domain of a dodecahedral graph, therefore, a solver must visit 20 vertices, without revisiting any of them. When such a trip makes a loop through all the vertices, or corners, of the graph, it is called a Hamiltonian cycle. When the first and last vertices in a trip are not connected, it is called a Hamiltonian path.

Fig. 5.2 An Icosian grid
(Wikimedia Commons)

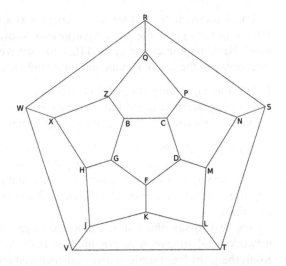

Exploring number-graph relations and how they combine into larger structures is a perfect example of blended (connective) cognition. Group theory is another case-in-point. As discussed several times, it is connected to several puzzles. The underlying question that group theory entertains dialectically is: How can we specify the members of a group so as to be able to combine them to make more members? In other words, by observing how numbers combine with each other, many new ideas related to structural possibilities might crystallize. An example of a group is the integers under the operation of addition, which has the following rule: any integer n plus another integer m always produces another integer. However, under division the integers do not form a group, because it is not true that an integer divided by another integer always produces a third. Some groups are infinite; others are finite, such as the binary digits under multiplication (Cayley 1854).

The Monty Hall Problem

Another famous puzzle that can be enlisted to take yet another look into the mathematical mind is the so-called *Monty Hall Problem* (MHP), named after television quiz show host Monty Hall who was the presenter of *Let's Make a Deal*, which premiered in 1963 and continued until the 2010s. It was formulated as a problem in probability theory by Steven Selvin in 1975. The contestants on the show had to select from three doors that hid different prizes. The situation can be broken down into stages as follows:

1. A contestant has to choose one of three doors: *A, B,* and *C.* Behind one is a new car, behind the other two are goats.
2. The contestant chooses one door, say *A.*
3. He or she has a ⅓ probability of having selected the car.

4. Monty Hall knows where the car is, so he says: "I'm not going to tell you what's behind door A, yet. But I will reveal that there is a goat behind door B."
5. Then he asks: "Will you now keep door A or swap to C?"

The assumption of most people is that the odds are 50-50 between A and C, so that switching would make no difference. But that is incorrect: C has a ⅔ probability of concealing the car, while A has just a ⅓ probability. This seems to be counterintuitive, but probability reasoning as we have seen several times already tends to reveal unexpected patterns. Elwes (2014: 334) explains the reasoning as follows:

> It may help to increase the number of doors, say, to 100. Suppose the contestant chooses door 54, with a 1% probability of finding the car. Monty then reveals that doors 1–53, 55–86, and 88–100 all contain wooden spoons. Should the contestant swap to 87, or stick with 54? The key point is that the probability that door 54 contains the car remains 1%, as Monty was careful not to reveal any information which affects this. The remaining 99%, instead of being dispersed around all the other doors, become concentrated at door 87. So she should certainly swap. The Monty Hall problem hinges on a subtlety. It is critical that Monty knows where the car is. If he doesn't, and opens one of the other doors at random (risking revealing the car but in fact finding a wooden spoon), then the probability has indeed shifted to ½. But in the original problem, he opens whichever of the two remaining doors he knows to contain a wooden spoon. And the contestant's initial probability of ⅓ is unaffected.

Of course, playing by the rules of probability may mean nothing if one loses—that is, finding oneself on a wrong point in the probability curve. However, knowing about the existence of the curve leads to many more insights into the nature of real events than common sense would seem to afford. The MHP reveals the uncanny ability of the mathematical imagination to unravel the hidden structure of real-life events. Our assumption that "two choices means 50–50 chances" is true when we know nothing about either choice. If we picked any coin then the chances of getting a head or tail are, of course, 50–50. But if we are told that one of the sides will come up more often because it has been somehow designed physically to do so, then everything changes. Indeed, information is what matters in computing outcomes.

In effect, the MHP brings out the principle that the more we know, the better our decision will be. If the number of doors in the MHP were 100, this becomes even clearer, as Elwes remarked. As Monty starts eliminating the bad candidates (in the 99 that were not chosen), he shifts the focus away from the bad doors to the good one more and more. After Monty's filtering, we are left with the original door and the other door. Here is where probability reasoning comes into play, allowing us to generalize the MHP—the probability of choosing the desired door improves as we get more information. Without any evidence, two choices are equally likely. As we gather additional evidence (and run more trials) we can increase our confidence interval:

1. Two choices are 50–50 when we know nothing about them.
2. Monty intervenes by "filtering" the bad choices on the other side.
3. In general, the more information, the more the possibility of re-evaluating our choices.

Logic Puzzles

The word *logic* has been used throughout this book, without providing any definition. In a sense, it is one part of the bi-part mind—the part that completes the work of
the imagination as has been argued throughout this book. It is the feature of the brain
that permits abstract generalizations. Defining it is nonetheless problematic because
it manifests itself in so many diverse ways. Perhaps the best way to explore its
features is through ingenious logic puzzles that require various features of logical
thinking to solve. One genre is due to Dudeney (1958). Here is a paraphrase of his
original puzzle:

> In a certain company, Bob, Janet, and Shirley hold the positions of director, engineer, and
> accountant, but not necessarily in that order. The accountant, who is an only child, earns the
> least. Shirley, who is married to Bob's brother, earns more than the engineer. What position
> does each person fill?

This simple puzzle, in its own miniature way, brings out what logical reasoning is
all about in a schematic, yet ingenious, way. We are told that: (1) the accountant is an
only child and (2) Bob has a brother (to whom, incidentally, Shirley is married). So,
from these two facts we can eliminate Bob as the accountant—since he is not an only
child. We are also told that the accountant earns the least of the three, and that Sheila
earns more than the engineer. From these facts, two obvious things about Sheila can
be established: (1) she is not the accountant (who earns the least, while she earns
more than someone else); (2) she is not the engineer (for she earns more than the
engineer). The only possibility for accountant left is Janet. Also by the process of
elimination, Sheila is the director and Bob the engineer.

As Dudeney's puzzle ingeniously showed, logical reasoning consists of several
processes working in tandem, such as elimination, deduction, and deriving unique
conclusions from specific facts. However, logical reasoning is not divorced from
abduction and its ability to connect the statements to reality or common sense. In
effect, Dudeney's puzzle shows that even in those puzzles that are classified as
purely logical, there is always some form of abduction at work. It really is a matter of
degree.

Perhaps no other puzzle genre exemplifies the use of logical reasoning more than
does Sudoku. Like Dudeney's puzzle, it involves trial and error, deduction, and an
occasional Aha insight. It is such a well-known pastime that it needs very little
explanation here. The basic puzzle is made up of a 9 × 9 grid, divided into 3 × 3
subgrids called boxes. The overall grid shows some numbers from 1 to 9 that have
been already inserted into their appropriate cells. The goal of the game is to fill in the
remaining empty cells, one number in each, so that each column, row, and box
contains all the numbers from 1 to 9, and so that no number is repeated in any row,
column, or inside a box. See Fig. 5.3.

Let's perform a few basic logical analyses for the sake of illustration. Looking at
the empty cell is in the top row of the grid we note that it is missing the number
8. Also the number 6 is missing from the right-most column. Now, looking at the
row starting with 7 (the fourth row from the top) we note that there are two empty

Fig. 5.3 A Sudoku puzzle

1	9	7	4	2	3		6	5
	4			6				9
	5		8		7	4	2	1
7	1		5	3	8	2		
	3	6		1			9	8
8				4		7	1	3
9		2	3		4	1	5	7
3	7		2			9		4
5	8	4	9		1		3	2

Fig. 5.4 Solution to the puzzle

1	9	7	4	2	3	8	6	5
2	4	8	1	6	5	3	7	9
6	5	3	8	9	7	4	2	1
7	1	9	5	3	8	2	4	6
4	3	6	7	1	2	5	9	8
8	2	5	6	4	9	7	1	3
9	6	2	3	8	4	1	5	7
3	7	1	2	5	6	9	8	4
5	8	4	9	7	1	6	3	2

cells, which must contain 4 and 9 in some order, since they are the two missing digits in the row, after having inserted the 6 at the end. If we put the 9 in the cell just before the inserted 6, we will produce a repetition of the 9 in the column that crosses the row. So, the 9 belongs to the third cell from the left in the row. We then can put the 4 in the remaining empty cell. Continuing to reason in this way, the diagram can be completed as shown in Fig. 5.4.

There are many variations to the puzzle format, with such colorful names as Killer Sudoku and Samurai Sudoku. As Nuessel (2013) elaborates, KenKen is the most mathematically useful derivative of the Sudoku concept. The first KenKen puzzles appeared in *The Times of London* and then in *The New York Times*. The instructions for doing a KenKen puzzle are as follows: the grid must be filled with digits that do not repeat in any row or column, and the digits in each heavily outlined box must produce the target number shown, by using addition, subtraction, multiplication, or division, as indicated in the box. A 4 × 4 grid uses the digits 1, 2, 3, 4, a 6 × 6 grid will use the digits 1, 2, 3, 4, 5, 6. The puzzle was invented by a Japanese mathematics instructor, Tetsuya Miyamoto, to teach his students arithmetic, logic, and patience.

In sum, Sudoku and its variants are based on the same type of archetype that characterizes most closed puzzles with an end-state—namely, a placement archetype, which is a derivative of the assembly one. In this case, the goal is to fill in cells according to rules so as to create a unique solution, much like the jigsaw and other assembly puzzles.

Some logic puzzles hide a trap, either because of the meaning of certain words, or because of the pseudo-mathematical style in which they present facts, creating a Gotcha Effect. The following classic puzzle never fails to confound solvers:

> Three women decide to go on a holiday to Las Vegas. They share a room at a hotel which is charging 1920s rates as a promotional gimmick. The women are charged only $10 each, or $30 in all. After going through his guest list, the manager discovers that he has made a mistake and has actually overcharged the three vacationers. The room the three are in costs only $25. So, he gives a bellhop $5 to return to them. The sneaky bellhop knows that he cannot divide $5 into three equal amounts. Therefore, he pockets $2 for himself and returns only $1 to each woman. Now, here's the conundrum. Each woman paid $10 originally and got back $1. So, in fact, each woman paid $9 for the room. The three of them together thus paid $9 × 3, or $27 in total. If we add this amount to the $2 that the bellhop dishonestly pocketed, we get a total of $29. Yet the women paid out $30 originally! Where is the other dollar?

The trap in this puzzle is not to be found in any single word, but in the way in which the numerical facts are laid out. Here is how the facts should be deconstructed in order to avoid the apparent discrepancy. Originally, the women paid out $30 for the room. That is how much money was in the hands of the hotel manager when he realized that he had overcharged them. He kept $25 of the $30, and gave $5 to the bellhop to return to the women. Each woman got back $1. This means that each one paid $9 for a room. Thus, altogether the three women spent $27. Of this money, the hotel got $25 and the other $2 was pilfered by the devious bellhop. So, there is no missing dollar.

Another way to explain this puzzle is as follows: We start by noting that the women paid out $30 for the room. Of this money, the manager kept $25. The women got back $3 ($1 each). So far this adds up to $25 + $3 = $28. The remaining $2 dollars was, of course, pocketed by the bellhop. Again, there is no missing dollar.

Mathematical Thinking

The discussion in this book was meant overall to indicate that puzzles are mirrors of the mathematical mind. The study of the mathematical mind has become a key area of inquiry within the cognitive sciences—an area of study now known as numerical or math cognition. It aims to determine how mathematics is learned and how math ideas emerge. Starting with the work of Stanislas Dehaene (1997), Brian Butterworth (1999), Keith Devlin (2000), and Lakoff and Núñez (2000), among others, such study has entailed an interdisciplinary approach criss-crossing fields such as psychology, philosophy, and anthropology. Today, there is a huge dataset of research findings and theories relating math cognition to math learning, and to how mathematics intersects with other neural faculties such as language and visual art. Above all else, the research suggests a certain frame of mind—which has been called blended (bi-part) thinking here—that undergirds the solution of puzzles and how some solutions can be generalized to produce mathematical knowledge. Although not stated explicitly as such in the relevant literature, the data on math cognition is consistent with the Generalization Principle of this book.

A particular puzzle, such as the Fibonacci Rabbit Puzzle, can be generalized as revealing the presence of a recursive infinite series within the solution, which, in turn, leads to the discovery of previously unknown math concepts. Having said this, it is true that math cognition is not easy to define psychologically, although we may have an intuitive sense of what it is. Generally, it is defined in two main ways. First, it is the awareness of quantity, space, and structural patterns among quantitative and spatial concepts. This definition is meant to reflect the possibility that math cognition may be innate. Second, it is defined as the awareness of how symbols stand for the concepts and how they encode them. At this level, math cognition is symbolic cognition and thus entails framing ideas through symbol-based artifacts such as puzzles that are, in effect, theories of both the mind and mathematics itself.

A historical starting point for the psychological study of math cognition—although not named in this way, of course—is the work of Immanuel Kant (2011: 278), who characterized mathematical thinking as the process of "combining and comparing given concepts of magnitudes, which are clear and certain, with a view to establishing what can be inferred from them." This precisely characterizes the abductive thinking process discussed throughout this book when solving puzzles. Kant argued further that the whole process becomes reflective and explicit when we examine the "visible signs" that we use to highlight the structural detail inherent in this type of knowledge. For example, a diagram of a triangle compared to that of a square will show where the differentiation occurs—one consists of three intersecting lines, while the other has four parallel and equal sides that form a boundary. This type of diagrammatic know-how is based on the brain's ability to synthesize scattered bits of information into holistic entities that can then be analyzed reflectively.

To see how the arguments made in this book fit in by and large with Kant's description, consider two ingenious puzzles. The first one is by Diophantus, which he formulated in his *Arithmetica:*

Divide a given number into two squares.

At first consideration, this seems to require a "brute force" approach embedded in trial and error. Now, while the trial-and-error method is sometimes part of deriving a solution, it is something that mathematicians wish to avoid in order to make the process efficient. Moreover, sometimes we cannot say for sure that trial-and-error can lead to a true generalizable solution. Mathematicians have always wanted to establish that something is the way it is beyond any shadow of a doubt; extrapolating a solution from given information by trial-and-error, on the other hand, can only suggest that something is the way it is, not demonstrate it. So, let's attack Diophantus' puzzle with a frame of mind that seeks to flesh out from it some hidden principle. The technique of setting an equation comes to mind to solve this in an easy and determinate fashion. The result is a quadratic equation, with the numbers being, 16, 16/5, and 12/5, so that $16 = (16/5)^2 + (12/5)^2$. The question becomes: What kind of thinking did he use? In this case, we actually have his solution in print, as reproduced below (cited in Bashmakova 1997: 24–25):

> Let the first summand be x^2, and thus the second, $16-x^2$. The latter is to be a square. I form the square of the difference of an arbitrary multiple x diminished by the root of 16, that is, diminished by 4. I form, for example, the square of $2x-4$. It is $4x^2 + 16-16x$. I put this expression equal to $16-x^2$. I add to both sides $x^2 + 16x$ and subtract 16. In this way I obtain $5x^2 = 16x$, hence $x = 16/5$. Thus one number is 256/25 and the other 144/25. The sum of these numbers is 16 and each summand is a square.

Diophantus' words reveal how his mind grappled with the puzzle. The Aha insight came from seeing the problem as a general one involving quadratic equations. This kind of thinking mirrors what Kant called transforming ideas into visible signs (in this case into a quadratic equation). The starting point is an Aha moment that leads to a visualization of the appropriate signs.

Now, let's consider the following puzzle devised by Lewis Carroll (1880: 7):

> A bag contains one counter, known to be either white or black. A white counter is put in, the bag shaken, and a counter drawn out, which proves to be white. What is now the chance of drawing a white counter?

To solve this, it is best to use the Kantian insight of transforming the information into visible signs, such as symbols. So, we let B and W-1 stand, respectively, for the black or white counter that might be in the bag at the start, and W-2 for the white counter added to the bag. The Aha insight is the fact that removing a white counter from the bag entails three equally likely combinations of two counters, one inside and one outside the bag:

Inside the Bag	Outside the Bag
(1) W-1	W-2
(2) W-2	W-1
(3) B	W-2

In combination (1), the white counter drawn out is the one that was put into the bag (W-2), and the white counter inside it (W-1) is the counter originally in it. Combination (2) is the converse of (1): the white counter drawn out is the one that was originally in the bag (W-1), and the white counter inside it (W-2) is the counter that was put in. In combination (3), the white counter drawn out is the one that was put in the bag (W-2), since there was no white counter originally in it. The counter that remains in the bag is a black one (B). In two of the three cases, Carroll observed, a white counter remains in the bag. So, the chance of drawing a white counter on the second draw is two out of three.

The relevant insight cannot be forced through a simple use of trial-and-error logic. It takes familiarity with all kinds of puzzles for a solution to "click in," in a manner of speaking. Carroll's puzzle zeroes in on Aha thinking in progress. It outlines the main features of the puzzle situation and then allows the mind to extrapolate a solution from them. In other words, it shows us that the parts relate to each other in specific ways.

The solution process is complete after the insight produced in this way is formalized. As discussed throughout this book, this suggests a flow model of how the mathematical mind moves from hunches to formalization. The hunch activates the imagination which, in turn, leads to some visual strategy or simply "visualization" (paraphrasing Kant), which leads to the Aha insight. This is then given some form (en equation, diagram, etc.) that suggests a simple solution. Finally, the experience of solving the puzzle can be given a logical form through generalization.

Puzzles are thus manifestations of the brain's innate sense of pattern that Brian Butterworth (1999) calls "numerosity." For Butterworth numbers constitute a separate and unique kind of intelligence with its own brain module, located in the left parietal lobe. But this alone does not guarantee that math know-how will emerge homogeneously in all individuals. Rather, the reason a person falters at math is not because of a "wrong gene" or "engine part" in the left parietal lobe, but because the individual has not fully developed the number sense with which he or she is born, and the reason is due to environmental and personal factors. It is no coincidence, therefore, that the left parietal lobe controls the movement of fingers, constituting a neurological clue to the evolution of our number sense, explaining why we count on our fingers. The seemingly nonlinguistic nature of math also might explain why cultures that have no symbols or words for numbers have still managed to develop counting systems for practical purposes.

There is a body of research that is supportive of Butterworth's basic thesis. In one study, Izard, Pica, Pelke, and Dehaene (2011) looked at notions of Euclidean geometry in an indigenous Amazonian society. The research team tested the hypothesis that certain aspects of non-perceptible Euclidean geometry map onto intuitions of space that are present in all humans (such as intuitions of points, lines, and surfaces), even in the absence of formal mathematical training. The Amazonian society is called the Mundurucu, and the subjects included adults and age-matched children controls from the United States and France as well as younger American children without education in geometry. The responses of Mundurucu adults and children coincided with those of mathematically educated adults and children, who were exposed formally to Euclidean geometry, and revealed an intuitive understanding of essential properties of such geometry. For instance, on a surface described to them as perfectly planar, the Mundurucu's estimations of the internal angles of triangles added up to approximately 180°, and when asked explicitly, they stated that there exists one single parallel line to any given line through a given point. These intuitions were also present in the group of younger American participants. The researchers concluded that, during childhood, humans develop geometrical intuitions that spontaneously accord with the principles of Euclidean geometry, even in the absence of training in such geometry.

The above finding could be explained otherwise—namely, that mathematical archetypes are at work here. If the research is correct, in fact, it would partially explain why we react to puzzles in parallel archetypal ways across the world. Keith Devlin's two books, titled *The Math Gene* (2000) and *The Math Instinct* (2005) also bring the notion of an innate number sense to the forefront, providing insights from relevant research and from reasoning about the kinds of cognitive structures that are involved in solving math puzzles. If there is some innate capacity for numerosity, which there must be, otherwise no one could count, why does it vary so widely, both among individuals in a specific culture and across cultures? The question is a key one. Devlin, unlike Butterworth, connects numerosity to language, since both are used by humans to model the world symbolically. As Devlin argues, our prehistoric ancestors' brains were essentially the same as ours, so they must have had the same underlying abilities. The difference is evolution—those brains could hardly have imagined how to multiply 15 by 36 or prove Fermat's Last Theorem. These are the result of historical forces at work in the imagination, leading to conceptual memes, as they have been called here, through the historical channel.

Stanislas Dehaene's *The Number Sense* (1997) is seen by many to be the key work that initiated the serious and systematic neuroscientific study of math cognition, bringing forth experimental evidence to suggest that human brains and those of some chimps come with a wired-in aptitude for math. The difference in the case of the latter is an inability to formalize this innate knowledge and then use it for invention and discovery. This is why certain ideas are found across cultures. Dehaene suggests that this rudimentary number sense is as basic to the way the brain understands the world as our perception of color or of objects in space, and, like these other abilities, our number sense is wired into the brain. Dehaene also shows that it was the invention of symbolic systems of numerals that started us on the climb to higher mathematics. He argues this by tracing the history of numbers, from early times when people indicated a number by pointing to a part of their body (even today, in some parts of New Guinea, the word for six is "wrist"), to early abstract numbers such as Roman numerals (chosen for the ease with which they could be carved into wooden sticks), to modern numerals and number systems. Dehaene also explores the unique and comparable abilities of idiot savants and mathematical geniuses, asking what might explain their special talents. Using modern imaging techniques (PET scans and fMRI), Dehaene reveals exactly where in the brain numerical calculation takes place. But perhaps most importantly, Dehaene argues that the human brain does not work like a computer, and that the physical world is not based on mathematics—rather, mathematics evolved to explain the physical world in a parallel way that the eye evolved to provide sight.

Perhaps the central topic in the study of math cognition is the following one: Is math related to language? The answer by Lakoff and Núñez (2000) is that the two share the same neuro-conceptual structures. Their basic claim is that the proofs and theorems of mathematics are arrived at, initially, through the same cognitive mechanisms that underlie language—analogy, metaphor, and metonymy. The idea that number sense stems from such "rhetorical thinking" certainly resonates with the ideas of Charles Peirce, who saw the source of all new ideas in the process of abduction, as discussed several times. But is this just speculation or is there some neuroscientific basis to it? Both in their 2000 book and in subsequent work, Lakoff and Núñez claim that this hypothesis can, and has, been substantiated with neurological techniques such as fMRI and other scanning devices, which has led them to adopt the notion of blending, whereby concepts in the brain that are sensed as "informing" each other become amalgams in a common neural substrate (Fauconnier and Turner 2002). Finding where this substrate is located and determining its characteristics is an ongoing goal of neuroscience.

Blending is used, for example, to explain the formation of negative numbers. These are derived from two basic conceptual metaphors, which Lakoff and Núñez call *grounding* and *linking*. Grounding metaphors encode basic ideas, being directly grounded in experience. For example, addition develops from the experience of counting objects and then inserting them in a collection. Linking metaphors connect concepts within mathematics that may or may not be based on physical experiences. Some examples of this are the number line, inequalities, and absolute value properties within an epsilon-delta proof of limit. Linking metaphors are the source of negative numbers, which emerge from a connective form of reasoning within the system of mathematics.

The underlying psychological hypothesis interweaved throughout this book is that both sides of the brain—the imaginative and the reflective (rational, logical) work in tandem, that is, in a blended fashion, to produce the new ideas inherent in the classical math puzzles. Even those that are designed to produce a Gotcha Effect are still ensconced in blended cognition. The same hypothesis applies to mathematical games of all kinds, even if they do not have the same dialectical structure of open puzzles (as discussed). Even a Sudoku puzzle requires that the solver imagine, or envision, number placements, combinations, and the like. When the diagram does not concede a definitive placement pattern, then we use hypothetical placements and from there make appropriate deductions. Thus, reflection or reasoning and imagination are blended even in such a seemingly straightforward game like Sudoku to produce relevant insights (Rosenhouse and Taalman 2011). If most puzzles are indeed solved first on the basis of imaginative Aha thinking, then it is relevant to think of this neurological capacity as a variable one and thus locatable on a scale, ranging from 0 or a low level of such thinking to 1, a maximum level.

So, any particular puzzle can be located on this hypothetical scale in terms of its degree of engagement of the imagination. The closer it is to "1," the more it requires activating it; the closer it is to "0," the less it is required and the more routine logical thinking is involved. The Nine-Dot Puzzle can clearly be located near the 1-point end, whereas many Sudoku puzzles fall nearer to the 0-end. Other kinds of puzzles fall somewhere in between. See Fig. 5.5.

The distinction between imaginative and reflective cognition corresponds to unconscious and conscious mental processes, respectively. The former is dominant during the processing of new information and the latter after an Aha insight has been

0 ━━━━━━━━━━━━━━━━━━━━━━━━━━━━━━ 1

Fig. 5.5 Aha thinking scale

achieved at which point the mind starts attempting to understand "what is going on," so to speak. In short, this means that puzzle-solving progresses through zones of understanding, from hunches to abstractions. The early zones are spontaneous and even emotionally based, as can be seen by the frustrations that an unsolved puzzle begets. The later stages are organizational and reflective.

The general psychological feature of this kind of blended thinking is its connective nature, whereby a situation that involves, say, crossing a river is then connected to other parallel situations via a puzzle's intuitive formulation. This suggests that the right hemisphere of the brain is where the process starts, leading subsequently to the left hemisphere's ability to encode and formalize the ideas grasped intuitively in the right hemisphere—a model that can be labeled "bimodal" (Danesi 2003). The essence of bimodality can be discerned in Howard Gardner's (1982: 74) statement: "Only when the brain's two hemispheres are working together can we appreciate the moral of a story, the meaning of a metaphor, words describing emotion, and the punch lines of jokes." It would seem that for the brain to interpret unfamiliar information it requires the experiential (probing) right-hemisphere functions to operate freely; these can be called R-Mode functions. However, both the anecdotal experiences of mathematicians and the neuroscientific findings also indicate that this exploratory effort would be virtually wasted if not followed up by the left hemisphere's analytical-reflective intervention; this capacity can be called L-Mode. All this suggests a general principle of puzzle-solving that has been called Cognitive Flow Theory in this book. It now can be formulated as follows: (1) the initial experiential-creative forms of puzzling involve R-Mode stages; (2) these produce the relevant insight, which is then given a more formal and analytical L-Mode treatment. Needless to say, a skilled mathematician who is already in firm control of the required L-Mode skills through previous understanding will not have to spend as much time on the R-Mode phase as would a neophyte. A consummate control of puzzling is, from a neurological perspective, a bimodal feat, requiring the integrated contribution of both the R-Mode and the L-Mode.

Supporting this model is a key study published in 1981 by Goldberg and Costa. The two neuroscientists suggested that the right hemisphere is a crucial point-of-departure for novel tasks because of its anatomical structure. Its greater connectivity with other centers in the complex neuronal pathways of the brain makes it a better "distributor" of new information. The left hemisphere, on the other hand, has a more sequentially organized neuronal-synaptic structure and, thus, finds it more difficult to assimilate information for which no previously formed synaptic circuits exist. This suggests that for any new input to be comprehensible, it must allow the synthetic functions of the R-Mode to do their work first.

At this point a caveat is in order. The claims made by those who subscribe to brain models of the mind, including the present author, are often products of a social trend. The "discovery" of the right hemisphere by neuroscience during the split-brain research era of the 1940s and 1950s was interpreted by many amateurs as vindicating the view that science and education were biased in favor of the left hemisphere, that is, in favor of analytical, deductive, and rational thought at the expense of creative, imaginative, and intuitive thinking. This led, shortly thereafter, to what Gardner

(1982: 266–267) aptly characterizes as "the temptation to tamper with the work of scientists," and consequently to a tendency to put forward sensational claims about "right-brain learning" by a coterie of "academic hucksters," as Gardner calls them. So, constructs derived from neuroscience, although highly plausible, are nonetheless subjective ones. Unless they lead to calculable results they are worthless in and of themselves. My sense is that bimodality is a theoretically viable one in the domain of puzzles; nevertheless, it still requires empirical research to substantiate it.

To summarize, in order to make something accessible to the L-Mode, it is necessary for the mind to explore the new structures and concepts embedded in a puzzle or game format through R-Mode thinking. This was evident in Alcuin's and Euler's puzzles. Once the initial R-Mode stage produced the relevant insight, the brain "shifted modes" and put the solver in a frame of mind to reflect on the new structural patterns *in themselves*—namely the principles of critical path theory, of graph theory, of combinatorics, and so on. It is this interplay between the two modes of thinking—the imaginative R-Mode and the analytical L-Mode—that is the likely source of the Aha experience. The suspense that accompanies an attempt to find a solution to a challenging puzzle, or the anxiety that develops from not finding one right away, is connected to the exploratory nature of the R-Mode and is thus a significant part of what makes puzzles so fascinating and engaging.

This model also fits in with the Generalization Principle formulated here, which seeks to explain the chain of discoveries that are triggered by a seemingly simple or trivial puzzle concept. So, the River-Crossing Puzzle became a math meme that spread subsequently throughout the world of mathematics leading to combinatorics, critical path theory, and other branches. One can still see the echocs of Alcuin's puzzle reverberating in these complex systems. The same can be said about Euler's Königsberg Bridges Puzzle, Fibonacci's Rabbit Puzzle, and so on. The primary objective of recreational mathematics is, arguably, to examine the thought processes that a puzzle activates so that a historical continuity between a complex system and its creative origins can be understood conceptually. As we saw with the *Ahmes Papyrus* and Alcuin's *Propositiones*, the ancient mathematicians understood that there is no better way to impart knowledge of mathematics than through puzzle formats. These stimulate interest and allow us to reconstruct or deconstruct the whole thinking process involved in making and solving them.

Another relevant psychological aspect of mathematical puzzles is that they often reveal archetypal thinking, explaining coincidences, such as the 2^{n-1} and 2^n-1 formulas that surface serendipitously in Kallikan's Chess Puzzle, Lucas's Tower of Hanoi Puzzle, and other puzzles. This implies that archetypal (unconscious) thinking might characterize the whole history of mathematical discovery, as argued throughout this book. Puzzles make complex ideas easy to understand, much like a riddle or a poem allows us to grasp more than what the words recount. Consider the concept of Markov Chains—a rather complex and difficult notion to understand without expert background knowledge. The notion can, however, be presented via a puzzle situation that encapsulates the probability laws describing the stochastic processes in the chains in a creative way, thus opening up the mind to its R-Mode grasp of the situation. Stochastic processes are those that have a random probability distribution or pattern, which may be analyzed statistically but may not be predicted precisely. A puzzle that encapsulates this notion brilliantly is called the *random walk*, introduced by George Polya in 1921:

Choose a point on a graph at the beginning. What is the probability that a random walker will reach it eventually? Or: What is the probability that the walker will return to his starting point?

Polya proved that the answer is 1, making it a virtual certainty, calling it a one-dimensional outcome. In higher dimensions this is not the case. A random walker on a three-dimensional lattice, for instance, has a much lower chance of returning to the starting point ($p = 0.34$). This leads to the notion of Markov Chain as a relevant model of the random walk, which is equivalent to flipping a coin to decide in which direction to go next. The defining characteristic of a Markov Chain is that the probability distribution at each stage depends only on the present, not the past. Markov Chains are thus perfect models for random walks and random events. A marker is placed at zero on the number line and a coin is flipped—if it lands on heads (H) the marker is moved one unit to the right (1); if it lands on tails (T), it is moved one unit to the left (-1). There are 10 ways of landing on 1 (by 3H and 2T), 10 ways of landing on -1 (2H and 3 T), 5 ways of landing on 3 (4H and 1 T), 5 ways of landing on -3 (1H and 4T), 1 way of landing on 5 (5H), and 1 way of landing on -5 (5 T).

Recall Ahmes' prologue to his *Papyrus* in which he hints at a deeper theory to puzzles. While this sense has become largely unconscious today, it nevertheless still underlies the common feeling of amazement that comes from unraveling the solution to a tricky puzzle or the development of a new perspective, which comes from reading a book such as *Flatland*. There is little doubt that the *Papyrus* was intended to intrigue its readers, to generate an Aha Effect, and thus to awe them with the power of numbers, mathematical symbols, and the relationships among them. As Ahmes aptly wrote, puzzle-solving seems to grant us "entrance into the knowledge of all existing things and all obscure secrets." As Henry Dudeney (cited by Wells 1992: 89) has also remarked: "The fact is that our lives are largely spent in solving puzzles; for what is a puzzle, but a perplexing question? And from childhood upwards we are perpetually asking questions and trying to answer them."

Ahmes' Legacy

Stuart Isacoff (2003) has argued that the invention of western musical traditions came about from the Pythagorean legacy that the "natural structure" of music is mathematical and mirrors the harmony of the emotions in humans. Thus, the "secret" as to why we react to, say, the symphonies of Beethoven so emotionally is due to its mathematical symmetry. Ian Stewart (2008: 9) has eloquently summarized this view as follows:

The main empirical support for the Pythagorean concept of a numerical universe comes from music, where they had noticed some remarkable connections between harmonious sounds and simple numerical facts. Using simple experiments they discovered that if a plucked string produces a note with a particular pitch, then a string half as long produces an extremely harmonious note, now called the octave. A string two-thirds as long produces the next most harmonious note, and one three-quarters as long also produces a harmonious note. These two numerical aspects of music are traced to the physics of vibrating strings, which move in patterns of waves. The number of waves that can fit into a given length of string is a whole number, and these whole numbers determine the simple numerical ratios. If the numbers do not form a simple ratio then the corresponding notes interfere with each other, forming discordant 'beats' which are unpleasant to the ear. The full story is more complex, involving what the brain is accustomed to, but there is a definite physical rationale behind the Pythagorean discovery.

Indirect support for Pythagoras' view came forward in 1865, when English chemist John Newlands arranged the elements according to atomic weight and discovered that those with similar properties occurred at every eighth element like the octaves of music. He called this finding, in fact, the Law of Octaves. In turn, it led to the development of the Periodic Law of chemical elements. But the Pythagorean Harmony of the Spheres view also revealed some unexpected and "unwanted" findings. When the sides of a right triangle are of unit length, then the hypotenuse turns out to be a non-integer value. This truly upset the Pythagoreans' harmonic view of the universe, as discussed briefly beforehand. Also, when ratios between certain string vibrations are set, other ratios are thrown off, thus producing dissonances. The Pythagoreans knew about these defects in their philosophy, but kept them secret. To banish the dissonances, the keyboard of the piano (clavier) was tempered by breaking the octave into equal parts, so that all harmonies sounded in tune—an event attributed to Johann Sebastian Bach's *Well-Tempered Clavier* (1742). It was an ingenious human invention that rectified the Pythagorean defect, thus establishing a musical tradition that continues to this day. This episode illustrates that discovery and invention may be two sides of the same coin. The difference is that discovery is serendipitous, whereas invention is intentional.

The Pythagorean approach to mathematics was, at its core, a search for intrinsic pattern in numbers, geometrical figures, and their manifestations in the universe. Within this framework, puzzles can be characterized as serving the Pythagorean ideal as structure-detecting devices. They flesh out structural principles in mathematically based ways as models of real-life situations. It actually takes genius to understand the obvious and reframe it in the form of a puzzle or a paradox. By introducing the relevant puzzles, paradoxes, and games into mathematics at specific points in its history the relevant theoretical offshoots have emerged. This has been called Ahmes' legacy throughout this book.

Of course, the connectivity among the different puzzles and their archetypes might often lead nowhere; but even so they reveal the bimodal or connective nature of cognition. Take, as an example, the following one, which is based on the ubiquitous Fibonacci sequence (Petkovic 2009: 15)

Make a rectangle without any gaps by using small squares whose sides are Fibonacci numbers.

Fig. 5.6 A solution to the
Fibonacci rectangle puzzle
(Wikimedia Commons)

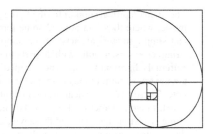

Imagining a solution is difficult. The question can be reworded, however, as drawing squares of sides using Fibonacci numbers and then assembling them with no spaces left over. One solution is shown in Fig. 5.6.

This puzzle possesses all the characteristics discussed in this book. First, it does not have an obvious solution, and thus requires insight thinking. It also connects the Fibonacci sequence to other areas of mathematics, since a spiral can be drawn serendipitously through it, as shown above. It thus involves archetypal thinking, since the spiral is, itself, an archetype that crops up in the golden ratio and in Nature. When all these features are compiled into an overall explanatory framework, it is obvious that this simple puzzle might be a miniature psychological model of the mathematical mind—a playful mind that is constantly involved in dialectic-ludic speculations such as this one.

As a final remark and illustration, it can be claimed that the blended mind is what makes mathematics so powerful. Consider one last time the Pythagorean theorem. The theorem generalizes the observation and subsequent proof that the square on the hypotenuse, c, is equal to the sum of the squares on the other two sides, a, b. As discussed, this was known previously by observation at a practical level; the theorem made it possible to represent this observation with a simple equation: $c^2 = a^2 + b^2$. It is at cognitive phases such as this one that, as Dehaene (1997: 151) puts it, "truth descends on" mathematicians, impelling them to imagine some new hypothesis, conjecture, or theory connected with the equation independently of its geometric referent. This exact kind of contemplation led, as is well known, to Fermat's Last Theorem. Pierre de Fermat was purportedly reading Diophantus' *Arithmetica* becoming keenly interested in Pythagorean triples—the sets of three numbers, a, b, and c, for which the equation $c^2 = a^2 = b^2$ is true. The triples were known far and wide before the advent of the theorem, even though they were not named as such (Neugebauer and Sachs 1945). Clay tablets recovered by archeologists, dating back to nearly 2000 BCE, reveal that the ancient Babylonians were familiar with many Pythagorean triples—3, 4, 5 ($3^2 + 4^2 = 5^2$); 6, 8, 10 ($6^2 + 8^2 = 10^2$); 5, 12, 13 ($5^2 + 12^2 = 13^2$); 8, 15, 17 ($8^2 + 15^2 = 17^2$); and so on. One clay tablet dating back to 1900 BCE, called *Plimpton 322*, contains fifteen numbered lines with two figures in each line that are Pythagorean.

Diophantus included a long discussion of the triples in his *Arithmetica*. In the margin of his copy of Diophantus' book, Fermat wrote the following enigmatic words (cited by Pappas 1991: 150):

> To divide a cube into two cubes, a fourth power, or in general any power whatever above the second, into two powers of the same denomination, is impossible, and I have assuredly found an admirable proof of this, but the margin is too narrow to hold it.

Fermat claimed (or thought) that his proof would show that only for the value $n = 2$ do solutions of $c^n = a^n + b^n$ exist: namely, for $c^2 = a^2 + b^2$. For four subsequent centuries, mathematicians across the world were intrigued by Fermat's claim, trying to come up with a proof, but always to no avail, although a number of special cases were settled. Gauss proved that $c^3 = a^3 + c^3$ had no positive solutions, and Fermat himself proved the untenability of $c^4 = a^4 + b^4$. Legendre gave a proof that $c^5 = a^5 + c^5$ had no solutions. And Dirichlet showed that $c^{14} = a^{14} + b^{14}$ had no solutions. But no general proof, as Fermat envisioned it, was discovered until in June 1993 mathematician Andrew Wiles declared that he had finally proved Fermat's Last Theorem. In December of that year, however, some mathematicians found a gap in his argument. In October of 1994 Wiles, together with Richard L. Taylor, filled that gap to virtually everyone's satisfaction. The Wiles-Taylor proof was published in May 1995 in the *Annals of Mathematics* (Wiles 1995, Taylor and Wiles 1995).

The proof is the result of connecting and modifying previous ideas and formulas used to explore the theorem. Two previous ideas, in fact, were crucial to it: the elliptic curve and the modular form. An elliptic curve is a curve of the form expressed by the following equation:

$$y^2 = x^3 + rx^2 + sx + t \text{ for integers } r, s, \text{ and } t$$

A modular form is a formula generalizing the Möbius transformation:

$$f(z) = (az + b)/(cz + d)$$

Two mathematicians, Yutaka Taniyama and Goro Shimura, hypothesized that every elliptic curve is associated with a modular form. It was quickly seen that a proof of the Taniyama-Shimura conjecture would imply the truth of Fermat's Last Theorem. In 1984 another mathematician, Gerhard Frey, saw that if Fermat's theorem were false—so the equation $a^p + b^p = c^p$ held for some positive integers a, b, c, and prime number p—then the elliptic curve $y^2 = x^3 + (b^p - a^p) x^2 - a^p b^p x$ would have such weird properties that it could not be modular. This would then contradict the Taniyama-Shimura conjecture. Wiles took Frey's idea and set out to prove the special case of the Taniyama-Shimura conjecture that implied the truth of Fermat's Last Theorem.

The gist of this whole discussion is, as Stewart (2012: 14) aptly puts it, that "Pythagoras's theorem, then, is important in its own right," since "it exerts even more influence through its generalizations." It has revealed the power of the human mind to infer things from previous ideas and then to apply them to discoveries and activities that would otherwise have never come to consciousness.

Fermat's Last Theorem still haunts some mathematicians, for the simple reason that the Wiles-Taylor proof was certainly not what Fermat could have envisioned. The proof required a computer program and depended on mathematical work subsequent to Fermat. In a pure sense, therefore, the Wiles-Taylor proof is not a historical resolution to Fermat's Last Theorem. Fermat left behind a true mathematical mystery. What possible "simple proof" could he have been thinking of as he read Diophantus' *Arithmetica*? As Stewart (1987: 48) aptly puts it, "Either Fermat was mistaken, or his idea was different." The point here is that without the Pythagorean theorem Fermat's Last Theorem and all the mathematical activities that it engendered would never have been possible. The equation $c^n = a^n + b^n$ impels mathematicians to think about a problem or a theorem in general, abstract ways rather than in specific, practical terms.

Connective thinking and modeling undergirds all discoveries in mathematics, at least in a generic sense. This is why one model leads to another and then to another and so on. Recall Pascal's triangle. It is a layout model that results from a triangular display of the expressions in the binomial expansion, $(a + b)^n$, revealing how certain numbers are related to others as well as how shapes and number patterns are intertwined. The numerical coefficients in the expansion form the shape of a triangle with infinite dimensions. This connection between a geometrical form, an algebraic concept, and the numerical coefficients involves connective thinking. It is by noticing the numbers in the triangular diagram that a further pattern emerges: Any number in a row is produced by the sum of the two numbers above it. Now, as we saw, Pascal's triangle crops up in many areas of mathematics. It turns up, for instance, in connection to the Fibonacci sequence. It also turns up in combinatorial analysis and in the structure of various curves.

Galileo's well-known assertion that the book of Nature is written in the language of mathematics alludes to the fact that mathematics is a descriptor and a predictor of natural phenomena. It was as well the view of Pythagoras, who saw a connectivity between mathematics, music, and the cosmos, encapsulating his view with the phrase "music of the spheres." Is mathematics truly the language of the spheres, allowing us to "discuss" the features of the universe in the form of equations, group theory, fractal diagrams, and the like? And if so, does it mean that mathematics is an investigative tool of the brain? As a corollary: Are puzzles the materials with which some parts of this tool are made?

In line with the Generalization Principle put forth in this book, a large portion of mathematics is built upon ingenious puzzle ideas, which continue to have many applications in science and computer programming. Discovery in science is guided and reflected by discovery and innovation in mathematics. The question is not, as Smolin (2013: 46) puts it, whether or not the mathematics is correct, but whether it is sufficient:

> Logic and mathematics capture aspects of nature, but never the whole of nature. There are aspects of the real universe that will never be representable in mathematics. One of them is that in the real world it is always some particular moment.

In conclusion, the mathematical mind cannot be studied in isolation, as can an object or a natural phenomenon. It cannot be taken out of the body for observation or inspection. By examining the products that this mind has produced we gain some indirect insights into its workings. Puzzles in particular shed light on some, if not most, of these very workings in a nutshell. It was Thomas Kuhn (1970) who coined the term "paradigm shift" to describe how progress in science occurs. That term applies perfectly to the case of puzzles, many of which have produced significant paradigm shifts in mathematics.

As Bronowski (1977: 24) has put it, to imagine means to make images and to move them about inside one's head in new arrangements:

> The images play out for us events which are not present to our senses, and thereby guard the past and create the future—a future that does not yet exist, and may never come to exist in that form. By contrast, the lack of symbolic ideas, or their rudimentary poverty, cuts off an animal from the past and the future alike, and imprisons it in the present. Of all the distinctions between man and animal, the characteristic gift which makes us human is the power to work with symbolic images.

The symbolic images, or archetypes, that puzzles enfold have influenced the course of a large part of mathematical history. To use Huygens' perfect expression one last time, it might well be that puzzles, when considered cumulatively, may harbor within them some "deep theory" that is embedded in the "obscure secrets," of which Ahmes spoke.

References

Abbott, Edwin (2002 [1884]). *The Annotated Flatland: A Romance of Many Dimensions*. Introduction and Notes by Ian Stewart. New York: Basic Books.

Adam, John A. (2004). *Mathematics in Nature: Modeling Patterns in the Natural World*. Princeton: Princeton University Press.

Alexander, James (2012). On the Cognitive and Semiotic Structure of Mathematics. In: M. Bockarova, M. Danesi, and R. Núñez (eds.), *Semiotic and Cognitive Science Essays on the Nature of Mathematics*, pp. 1–34. Munich: Lincom Europa.

Al-Khalili, Jim (2012). *Paradox: The Nine Greatest Enigmas in Physics*. New York: Broadway.

Andrews, William S. (1960). *Magic Squares and Cubes*. New York: Dover.

Auble, Pamela, Franks, Jeffrey and Soraci, Salvatore (1979). Effort Toward Comprehension: Elaboration or Aha !? *Memory & Cognition* 7: 426–434.

Ascher, Marcia (1990). A River-Crossing Problem in Cross Cultural Perspective. *Mathematics Magazine* 63: 26–29.

Averbach, Bonnie and Chein, Orin (1980). *Problem Solving Through Recreational Mathematics*. New York: Dover.

Bachet, Claude-Gaspar (1984). *Problèmes plaisans et délectables qui se font par les nombres*. Lyon: Gauthier-Villars.

Ball, W. W. Rouse (1972). *Mathematical Recreations and Essays*, 12th edition, revised by H. S. M. Coxeter. Toronto: University of Toronto Press.

Banks, Robert S. (1999). *Slicing Pizzas, Racing Turtles, and Further Adventures in Applied Mathematics*. Princeton: Princeton University Press.

Barwise, Jon and Etchemendy, John (1986). *The Liar*. Oxford: Oxford University Press.

Bashmakova, Isabella G. (1997). *Diophantus and Diophantine Equations*. Washington, D.C.: Mathematical Association of America.

Basin, S. L. (1963). The Fibonacci Sequence as It Appears in Nature. *The Fibonacci Quarterly*, 1 (1963), 53–64.

Benjamin, Arthur, Chartrand, Gary, and Zhang, Ping (2015). *The Fascinating World of Graph Theory*. Princeton: Princeton University Press.

Benson, Donald C. (1999). *The Moment of Proof: Mathematical Epiphanies*. Oxford: Oxford University Press.

Bennett, G. T. (1910). The Eight Queens Problem. *Messenger of Mathematics* 39: 19.

Bergin, Thomas G. and Fisch, Max (eds. and trans.) (1984). *The New Science of Giambattista Vico*. Ithaca: Cornell University Press.

Berlinski, David (2013). *The King of Infinite Space: Euclid and His Elements*. New York: Basic Books.

© Springer International Publishing AG, part of Springer Nature 2018
M. Danesi, *Ahmes' Legacy*, Mathematics in Mind,
https://doi.org/10.1007/978-3-319-93254-5

Biggs, N. L. (1979). The Roots of Combinatorics. *Historia Mathematica* 6: 109-136.

Bohning, Gerry and Althouse, Jody K. (1997). Using Tangrams to Teach Geometry to Young Children. *Early Childhood Education Journal* 24: 239–242.

Bor, Daniel (2012). *The Ravenous Brain: How the New Science of Consciousness Explains Our Insatiable Search for Meaning.* New York: Basic Books.

Borovkov, Alexander A. (2013). *Probability Theory.* New York: Springer.

Bronowski, Jacob (1973). *The Ascent of Man.* Boston: Little, Brown, and Co.

Bronowski, Jacob (1977). *A Sense of the Future.* Cambridge: MIT Press.

Brooke, Maxey (1969). *150 Puzzles in Crypt-Arithmetic.* New York: Dover.

Bruno, Giuseppe, Genovese, Andrea, and Improta, Gennaro (2013). Routing Problems: A Historical Perspective. In: M. Pitici (ed.), *The Best Writing in Mathematics 2012.* Princeton: Princeton University Press.

Burkholder, Peter (1993). Alcuin of York's *Propositiones ad acuendos juvenes:* Introduction, Commentary & Translation. *History of Science & Technology Bulletin,* Vol. 1, number 2.

Butterworth, Brian (1999). *What Counts: How Every Brain is Hardwired for Math.* Michigan: Free Press.

Cantor, Georg (1874). Über eine Eigenschaft des Inbegriffes aller reelen algebraischen Zahlen. *Journal für die Reine und Angewandte Mathematik* 77: 258–262.

Cardano, Girolamo (1663 [1961]). *The Book on Games of Chance (Liber de ludo aleae).* New York: Holt, Rinehart, and Winston.

Carl Sagan (1985). *Contact.* New York: Pocket Books.

Carroll, Lewis (1860). *A Syllabus of Plane Algebraical Geometry.* Oxford University Notes.

Carroll, Lewis (1879). *Euclid and His Modern Rivals.* London: Macmillan.

Carroll, Lewis (1880). *Pillow Problems and a Tangled Tale.* New York: Dover.

Carroll, Lewis (1958a). *The Game of Logic.* New York: Dover.

Carroll, Lewis (1958b). *Mathematical Recreations of Lewis Carroll.* New York: Dover.

Cayley, Arthur (1854). On the Theory of Groups, as Depending on the Symbolic Equation $\theta^n = 1$. *Philosophical Magazine* 7: 40–47

Chaitin, Gregory J. (2006). *Meta Math.* New York: Vintage.

Changeux, Pierre (2013). *The Good, the True, and the Beautiful: A Neuronal Approach.* New Haven: Yale University Press.

Chase, Arnold B. (1979). *The Rhind Mathematical Papyrus: Free Translation and Commentary with Selected Photographs, Transcriptions, Transliterations and Literal Translations.* Reston, VA: National Council of Teachers of Mathematics.

Clark, Michael (2012). *Paradoxes from A to Z.* London: Routledge.

Conrad, Axel, Hindrichs, Tanja, Morsy, Hussein, and Wegener, Ingo (1994). Solution of the Knight's Hamiltonian Path Problem on Chessboards. *Discrete Applied Mathematics* 50: 125–134.

Conway, John Horton (2000). *On Numbers and Games.* Natick, Mass.: A. K. Peters.

Cook, William J. (2014). *In Pursuit of the Traveling Salesman Problem.* Princeton: Princeton University Press.

Costello, Matthew J. (1988). *The Greatest Puzzles of All Time.* New York: Dover.

Crilly, Tony (2011). *Mathematics.* London: Quercus.

Cutler, Bill (2003). Solution to Archimedes' Loculus. http://www.billcutlerpuzzles.com.

Dalgety, James and Hordern, Edward (1999). Classification of Mechanical Puzzles and Physical Objects Related to Puzzles. In: Elwyn Berlekamp and Tom Rodgers (eds.), *The Mathemagician and Pied Puzzler: A Collection in Tribute to Martin Gardner,* pp. 175–186. Natick, Mass.: A. K. Peters.

Danesi, Marcel (2002). *The Puzzle Instinct: The Meaning of Puzzles in Human Life.* Bloomington: Indiana University Press.

Danesi, Marcel (2003). *Second Language Teaching: A View from the Right Side of the Brain.* Dordrecht: Kluwer.

Danesi, Marcel (2016). *Language and Mathematics: In Interdisciplinary Guide.* Berlin: Mouton de Gruyter.

Danforth, Samuel (1647). *MDCXLVII, an Almanac for the Year of Our Lord 1647*. ProQuest 2011.

Darling, David (2004). *The Universal Book of Mathematics: From Abracadabra to Zeno's Paradoxes*. New York: John Wiley and Sons.

Davis, Paul J. and Hersh, Reuben (1986). *Descartes' Dream: The World according to Mathematics*. Boston: Houghton Mifflin.

Dawkins, Richard (1976). *The Selfish Gene*. Oxford: Oxford University Press.

De Bono, Edward (1970). *Lateral Thinking: Creativity Step-by-Step*. New York: Harper & Row.

De Grazia, Joseph (1981). *Math Tricks, Brain Twisters & Puzzles*. New York: Bell Publishing Company.

Dehaene, Stanislas (1997). *The Number Sense: How the Mind Creates Mathematics*. Oxford: Oxford University Press.

Dehaene, Stanislas (2014). *Consciousness and the Brain*. New York: Penguin Books.

De Morgan, Augustus (1872). *A Budget of Paradoxes*. Library of Alexandria.

Derbyshire, John (2004). *Prime Obsession: Bernhard Riemann and His Greatest Unsolved Problem in Mathematics*. Washington: Joseph Henry Press.

Devlin, Keith (2000). *The Math Gene: How Mathematical Thinking Evolved and Why Numbers Are Like Gossip*. New York: Basic.

Devlin, Keith (2005). *The Math Instinct*. New York: Thunder's Mouth Press.

Devlin, Keith (2011). *The Man of Numbers: Fibonacci's Arithmetic Revolution*. New York: Walker and Company.

Dorrie, Heinrich (1965). *100 Great Problems in Elementary Mathematics*. New York: Dover.

Dudeney, Henry E. (1917). *Amusements in Mathematics*. New York: Dover.

Dudeney, Henry E. (1958). *The Canterbury Puzzles and Other Curious Problems*. New York: Dover.

Dudeney, Henry E. (2016, reprint). *536 Puzzles and Curious Problems*. New York: Dover.

Dunlap, Richard A. (1997). *The Golden Ratio and Fibonacci Numbers*. Singapore. World Scientific.

Du Sautoy, Marcus (2004). *The Music of the Primes: Bernhard Riemann and the Greatest Unsolved Problem in Mathematics*. New York: HarperCollins.

Eco, Umberto (1983). *The Name of the Rose* New York: Picador.

Eco, Umberto (1989). *The Open Work*. Cambridge: Harvard University Press.

Eco, Umberto (1998). *Serendipities: Language and Lunacy*, translated by William Weaver. New York: Columbia University Press.

Elwes, Richard (2014). *Mathematics 1001*. Buffalo: Firefly.

Erdös, Paul (1934). A Theorem of Sylvester and Schur. *Journal of the London Mathematical Society* 9: 282–288.

Euclid (1956). *The Thirteen Books of Euclid's Elements*, 3 Volumes. New York: Dover.

Fauconnier, Gilles and Turner, Mark (2002). *The Way We Think: Conceptual Blending and the Mind's Hidden Complexities*. New York: Basic.

Fibonacci, Leonardo (2002). *Liber Abaci*, trans. by L. E. Sigler. New York: Springer.

Flood, Robert and Wilson, Raymond (2011). *The Great Mathematicians: Unravelling the Mysteries of the Universe*. London: Arcturus.

Fortnow, Lance (2013). *The Golden Ticket: P, NP, and the Search for the Impossible*. Princeton: Princeton University Press.

Foulds, L. R. and Johnston, D. G. (1984). An Application of Graph Theory and Integer Programming: Chessboard Non-Attacking Puzzles. *Mathematics Magazine* 57: 95–104.

Frege, Gottlob (1879). *Begiffsschrift eine der aritmetischen nachgebildete Formelsprache des reinen Denkens*. Halle: Nebert.

Freiberger, Marianne (2006). *Flatland: A Review. Plus*, https://plus.maths.org/content/flatland.

Frey, Alexander H. and Singmaster, David (1982). *Handbook of Cubik Math*. Hillside, N.J.: Enslow.

Gale, David (1979). The Game of Hex and the Brouwer Fixed-Point Theorem. *The American Mathematical Monthly* 86: 818–827.

Gardner, Martin (1970). The Fantastic Combinations of John Conway's New Solitaire Game "Life". *Scientific American* 223: 120–123.

Gardner, Martin (1979a). Chess Problems on a Higher Plane, Including Images, Rotations and the Superqueen. *Scientific American* 240: 18–22.

Gardner, Martin (1979b). *Aha! Insight!* New York: Scientific American.

Gardner, Martin (1982). *Gotcha! Paradoxes to Puzzle and Delight.* San Francisco: Freeman.

Gardner, Martin (1994). *My Best Mathematical and Logic Puzzles.* New York: Dover.

Gardner, Martin (1997). *The Last Recreations: Hydras, Eggs, and Other Mathematical Mystifications.* New York: Copernicus.

Gardner, Martin (1998). A Quarter-Century of Recreational Mathematics. *Scientific American* 279: 68–75.

Gerdes, Paulus (1994). On Mathematics in the History of Sub-Saharan Africa *Historia Mathematica* 21: 23–45.

Gessen, Masha (2009). *Perfect Rigor: A Genius and the Mathematical Breakthrough of the Century.* Boston: Houghton Mifflin Harcourt.

Gillings, Richard J. (1961). Think-of-a-Number: Problems 28 and 29 of the Rhind Mathematical Papyrus. *The Mathematics Teacher* 54: 97–102.

Gillings, Richard J. (1962). Problems 1 to 6 of the Rhind Mathematical Papyrus. *The Mathematics Teacher* 55: 61–65.

Gillings, Richard J. (1972). *Mathematics in the Time of the Pharaohs.* Cambridge, Mass.: MIT Press.

Gödel, Kurt (1931). Über formal unentscheidbare Sätze der Principia Mathematica und verwandter Systeme, Teil I. *Monatshefte für Mathematik und Physik* 38: 173–189.

Goldberg, E. and Costa, L. D. (1981). Hemispheric Differences in the Acquisition of Descriptive Systems. *Brain and Language* 14: 144–173.

Gosset, T. (1914). The Eight Queens Problem. *Messenger of Mathematics* 44: 48.

Hadley, John and Singmaster, David (1992). Problems to Sharpen the Young. *Mathematics Gazette* 76: 102–126.

Haken, Wolfgang (1977). Every Planar Map Is Four-Colorable. *Illinois Journal of Mathematics* 21: 429–567.

Haken, Wolfgang and Appel, Kenneth (1977). The Solution of the Four-Color-Map Problem. *Scientific American* 237: 108–121.

Haken, Wolfgang and Appel, Kenneth (2002). The Four-Color Problem. In: D. Jacquette (ed.), *Philosophy of Mathematics*, pp. 193-208. Oxford: Blackwell.

Hales, Thomas C. (1994). The Status of the Kepler Conjecture. *The Mathematical Intelligencer* 16: 47–58.

Hales, Thomas C. (2000). Cannonballs and Honeycombs. *Notices of the American Mathematical Society* 47: 440-449.

Hales, Thomas C. (2005). A Proof of the Kepler Conjecture. *Annals of Mathematics. Second Series* 62: 1065–1185.

Hales, Thomas C. and Ferguson, Samuel P. (2006). A Formulation of the Kepler Conjecture. *Discrete & Computational Geometry* 36: 21–69.

Hales, Thomas C. and Ferguson, Samuel P. (2011). *The Kepler Conjecture: The Hales-Ferguson Proof.* New York: Springer.

Hannas, Linda (1972). *The English Jigsaw Puzzle, 1760–1890.* London: Wayland.

Hannas, Linda (1981). *The Jigsaw Book: Celebrating Two Centuries of Jigsaw-Puzzling Round the World.* New York: Dial.

Hayan Ayaz, Izzetoglu, Meltem, Shewokis, Patricia, and Onaral, Banu (2012). Tangram Solved? Prefrontal Cortex Activation Analysis during Geometric Problem Solving. *IEEE Conference Proceedings*: 4724–4727.

Heath, Thomas L. (1958). *The Works of Archimedes with the Method of Archimedes.* New York: Dover.

Hellman, Hal (2006). *Great Feuds in Mathematics: Ten of the Liveliest Disputes Ever.* Hoboken: John Wiley.

Hersh, Reuben (1998). *What Is Mathematics, Really?* Oxford: Oxford University Press.

Hersh, Reuben (2014). *Experiencing Mathematics.* Washington, DC: American Mathematical Society.

Hesse, Hermann (1943). *Magister Ludi* (New York: Bantam).

van Hiele, Pierre M. (1984). The Child's Thought and Geometry. In: David Fuys, Dorothy Geddes, and Rosamond Tischler (eds.), *English Translations of Selected Writings of Dina van Hiele-Geldof and P. M. van Hiele*, pp. 243–252. Brooklyn: Brooklyn College of Education.

Hilbert, David (1931). Die Grundlagen der elementaren Zahlentheorie. *Mathematische Annalen* 104: 485–494.

Hinz, Andreas M., Klavzar, Sandi, Milutinovic, Uros, and Petr, Ciril (2013). *The Tower of Hanoi: Myths and Maths.* Basel: Birkhaüser.

Hofstadter, Douglas (1979). *Gödel, Escher, Bach: An Eternal Golden Braid.* New York: Basic Books.

Hofstadter, Douglas and Sander, Emmanuel (2013). *Surfaces and Essences: Analogy as the Fuel and Fire of Thinking.* New York: Basic Books.

Hooper, William (1782). *Rational Recreations.* London: L. Davis.

Hovanec, Helene (1978). *The Puzzlers' Paradise: From the Garden of Eden to the Computer Age.* New York: Paddington Press.

Hudson, Derek (1954). *Lewis Carroll: An Illustrated Biography.* London: Constable.

Huizinga, Johan (1938). *Homo Ludens: A Study of the Play-Element in Human Culture.* New York: Beacon Press.

Hunter, J. A. H. (1965). *Fun with Figures.* New York: Dover.

Isacoff, Stuart (2003). *Temperament: How Music Became a Battleground for the Great Minds of Western Civilization.* New York: Knopf.

Izard, Veronique, Pica, Pierre, Pelke, Elizabeth S., and Dehaene, Stephen (2011). Flexible Intuitions of Euclidean geometry in an Amazonian Indigene Group. *PNAS* 108: 9782–9787.

Jung, Carl G. (1983). *The Essential Jung.* Princeton: Princeton University Press.

Kant, Immanuel (2011 [1910]). *Critique of Pure Reason*, trans. J. M. D. Meiklejohn. CreateSpace Platform.

Kasner, Edward and Newman, James R. (1940). *Mathematics and the Imagination.* New York: Simon and Schuster.

Kershaw, Trina and Ohlsson, Stellan (2004). Multiple Causes of Difficulty in Insight: The Case of the Nine-Dot Problem. *Journal of Experimental Psychology: Learning, Memory, and Cognition* 30: 3–13.

Kim, Scott (2016). *What Is a Puzzle?* scottkim.com.previewc40.carrierzone.com/thinkinggames/whatisapuzzle.

Klarner, David A. (1967). Cell Growth Problems. *Canadian Journal of Mathematics* 19: 851-863.

Klarner, David A. (ed.) (1981). *Mathematical Recreations: A Collection in Honor of Martin Gardner.* New York: Dover.

Kline, Morris (1985). *Mathematics and the Search for Knowledge.* Oxford: Oxford University Press.

Knuth, Donald E. (1974). *Surreal Numbers.* Boston: Addison-Wesley.

Kosslyn, Stephen M. (1983). *Ghosts in the Mind's Machine: Creating and Using Images in the Brain.* New York: W. W. Norton.

Kosslyn, Stephen M. (1994). *Image and Brain.* Cambridge, Mass.: MIT Press.

Kraitchik, Maurice (1942). *Mathematical Recreations.* New York: W. W. Norton.

Kuhn, Thomas S. (1970). *The Structure of Scientific Revolutions.* Chicago: University of Chicago Press.

Kurzweil, Ray (2012). *How to Create a Mind: The Secret of Human Thought Revealed.* New York: Viking.

Lakoff, George and Núñez, Rafael (2000). *Where Mathematics Comes From: How the Embodied Mind Brings Mathematics into Being*. New York: Basic Books.

Li, Yan and Du, Shiran (1987). *Chinese Mathematics: A Concise History*, translated by J. H. Crossley and A. W-C. Lun. Oxford: Oxford University Press.

Livio, Mario (2002). *The Golden Ratio: The Story of Phi, the World's Most Astonishing Number*. New York: Broadway Books.

Loyd, Sam (1914). *Cyclopedia of Tricks and Puzzles*. New York: Dover.

Loyd, Sam (1952). *The Eighth Book of Tan*. New York: Dover.

Loyd, Sam (1959–1960). *Mathematical Puzzles of Sam Loyd*, 2 volumes, compiled by Martin Gardner. New York: Dover.

Lucas, Edouard A. (1882–1894). *Récreations mathématiques*, 4 vols. Paris: Gauthier-Villars.

Martin, Robert (2004). The St. Petersburg Paradox. In: *The Stanford Encyclopedia*. Stanford: Stanford University Press.

Maor, Eli (1998). *Trigonometric Delights*. Princeton: Princeton University Press.

Margalit, Avishai and Bar-Hillel, M. (1983). Expecting the Unexpected. *Philosophia* 13: 337–344.

Matthews, William H. (1970). *Mazes & Labyrinths: Their History & Development*. New York: Dover.

Merton, Robert K. and Barber, Elinor (2003). *The Travels and Adventures of Serendipity: A Study in Sociological Semantics and the Sociology of Science*. Princeton: Princeton University Press.

Morrow, Glenn R. (1970). *A Commentary on the First Book of Euclid's Elements*. Princeton: Princeton University Press

Nave, Ophir, Neuman, Yair, Howard, Newton, and Perslovsky, L. (2014). How Much Information Should We Drop to Become Intelligent? *Applied Mathematics and Computation* 245: 261–264.

Netz, Reviel and Noel, William (2007). *The Archimedes Codex: Revealing the Secrets of the World's Greatest Palimpsest*. London: Weidenfeld & Nicholson.

Neugebauer, Otto and Sachs, Joseph (1945). *Mathematical Cuneiform Texts*. New Haven: American Oriental Society.

Neumann, John von (1958). *The Computer and the Brain*. New Haven: Yale University Press.

Neuman, Yair (2007). Immune Memory, Immune Oblivion: A Lesson from Funes the Memorious. *Progress in Biophysics and Molecular Biology* 92: 258267.

Neuman, Yair (2014). *Introduction to Computational Cultural Psychology*. Cambridge: Cambridge University Press.

Nuessel, Frank (2013). The Representation of Mathematics in the Media. In: M. Bockarova, M. Danesi and R. Núñez (eds.), *Semiotic and Cognitive Science Essays on the Nature of Mathematics*, pp. 154-198. Munich: Lincom Europa.

Northrop, Eugene (1944). *Riddles in Mathematics*. London: Penguin.

Ogilvy, C. (1956). *Excursions in Mathematics*. New York: Dover.

Olivastro, Dominic (1993). *Ancient Puzzles: Classic Brainteasers and Other Timeless Mathematical Games of the Last 10 Centuries*. New York: Bantam.

O'Shea, Donal (2007). *The Poincaré Conjecture*. New York: Walker.

Pappas, Theoni (1991). *More Joy of Mathematics*. San Carlos: Wide World Publishing.

Peet, Thomas E. (1923). *The Rhind Papyrus*. Liverpool: University of Liverpool Press.

Peirce, Charles S. (1931–1958). *Collected Papers of Charles Sanders Peirce*, ed. by C. Hartshorne, P. Weiss and A.W. Burks, vols. 1-8. Cambridge: Harvard University Press.

Petkovic, Miodrag S. (2009). *Famous Puzzles of Great Mathematicians*. Providence, RI: American Mathematical Society.

Phillips, Hubert (1966). *Caliban's Problem Book*. New York: Dover.

Plato (2004). *The Republic*, ed. by C. D. C. Reeve. Indianapolis: Hackett.

Plato (2006). *Meno*, ed. by Dominic Scott. Cambridge: Cambridge University Press.

Pohl, Ira (1967). A Method for Finding Hamiltonian Paths and Knight's Tours. *Communications of the ACM* 10(7): 446–449.

Polya, George (1921). Über eine Aufgabe der Wahrscheinlichkeitsrechnung betreffend die Irrfahrt im Strassennetz. *Mathematische Annalen* 84: 149–160.

Posamentier, Alfred S. and Lehmann, Ingmar (2007). *The (Fabulous) Fibonacci Numbers*. Amherst: Prometheus.

Pressman, Ian and Singmaster, David (1989). The Jealous Husbands and the Missionaries and Cannibals. *The Mathematical Gazette* 73: 73-81.

Read, Ronald C. (1965). *Tangrams: 330 Tangram Puzzles* New York: Dover.

Richards, Dana (1999). Martin Gardner: A "Documentary." In: E. Berlekamp and T. Rodgers (eds.), *The Mathemagician and Pied Puzzler: A Collection in Tribute to Martin Gardner*, pp. 3-12. Natick, Mass.: A. K. Peters.

Richeson, David S. (2008). *Euler's Gem: The Polyhedron Formula and the Birth of Topology*. Princeton: Princeton University Press.

Rockmore, Dan (2005). *Stalking the Riemann Hypothesis: The Quest to Find the Hidden Law of Prime Numbers*. New York: Vintage.

Roberts, Royston M. (1989). *Serendipity: Accidental Discoveries in Science*. New York: John Wiley.

Robins, R. Gay and Shute, Charles C. D. (1987). *The Rhind Mathematical Papyrus: An Ancient Egyptian Text*. London: British Museum Publications Limited.

Rosenhouse, Jason and Taalman, Laura (2011). *Taking Sudoku Seriously*. Oxford: Oxford University Press.

Rucker, Rudy (1987). *Mind Tools: The Five Levels of Mathematical Reality*. Boston: Houghton Mifflin.

Russell, Bertrand (1918). *The Philosophy of Logical Atomism*. London: Routledge.

Russell, Bertrand and Whitehead, Alfred North (1913). *Principia mathematica*. Cambridge: Cambridge University Press.

Sabbagh, Karl (2004). *The Riemann Hypothesis: The Greatest Unsolved Problem in Mathematics*. New York: Farrar, Strauss & Giroux.

Schneider, Michael S. (1994). *Constructing the Universe: The Mathematical Archetypes of Nature, Art, and Science*. New York: Harper Collins.

Schuh, Fred (1968). *The Master Book of Mathematical Recreations*. New York: Dover.

Schwenk, Allen J. (1991). Which Rectangular Chessboards Have a Knight's Tour? *Mathematics Magazine* 64: 325–332.

Shapiro, Stuart C. (1998). A Procedural Solution to the Unexpected Hanging and Sorites Paradoxes. *Mind* 107: 751–761.

Selvin, Steven (1975). A Problem in Probability (letter to the editor). *American Statistician* 29: 67.

Singmaster, David (1998). The History of Some of Alcuin's *Propositiones*. In: P. L. Butzer, H. Th. Jongen, and W. Oberschelp (eds.), *Charlemagne and His Heritage: 1200 Years of Civilization and Science in Europe, Vol. 2*, pp. 11–29. Brepols: Turnhout.

Slocum, Jerry and Botermans, Jack (1994). *The Book of Ingenious and Diabolical Puzzles*. New York: Times Books.

Smolin, Lee (2013). *Time Reborn: From the Crisis in Physics to the Future of the Universe*. Boston: Houghton Mifflin Harcourt.

Smullyan, Raymond (1978). *What Is the Name of this Book? The Riddle of Dracula and Other Logical Puzzles*. Englewood Cliffs, N.J.: Prentice-Hall.

Smullyan, Raymond (1979). *The Chess Mysteries of Sherlock Holmes*. New York: Knopf.

Smullyan, Raymond (1982). *Alice in Puzzle-Land*. Harmondsworth: Penguin.

Smullyan, Raymond (1997). *The Riddle of Scheherazade and Other Amazing Puzzles, Ancient and Modern*. New York: Knopf.

Spalinger, Anthony (1990). The Rhind Mathematical Papyrus as a Historical Document. *Studien zur Altägyptischen Kultur* 17: 295–337.

Sternberg, Robert J. (1985). *Beyond IQ: A Triarchic Theory of Human Intelligence*. New York: Cambridge University Press.

Stewart, Ian (1987). *From Here to Infinity: A Guide to Today's Mathematics.* Oxford: Oxford University Press.

Stewart, Ian (2008). *Taming the Infinite.* London: Quercus.

Stewart, Ian (2012). *In Pursuit of the Unknown: 17 Equations That Changed the World.* New York: Basic Books.

Strohmeier, John and Westbrook, Peter (1999). *Divine Harmony: The Life and Teachings of Pythagoras.* Berkeley, CA: Berkeley Hills Books.

Swetz, Frank J. and Kao, T. I. 1977. *Was Pythagoras Chinese? An Examination of Right-Triangle Theory in Ancient China.* University Park: Pennsylvania State University Press.

Takagi, Shigeo (1999). Japanese Tangram: The Sei Shonagon Pieces. In: E. Berlekamp and T. Rodgers, (eds.), *The Mathemagician and Pied Puzzler: A Collection in Tribute to Martin Gardner,* pp. 97–98. Natick, Mass.: A. K. Peters.

Tall, David (2013). *How Humans Learn to Think Mathematically.* Cambridge: Cambridge University Press.

Tarski, Alfred. (1933 [1983]). *Logic, Semantics, Metamathematics, Papers from 1923 to 1938,* John Corcoran (ed.). Indianapolis: Hackett Publishing Company.

Taylor, Richard and Wiles, Andrew (1995). Ring-Theoretic Properties of Certain Hecke Algebras. *Annals of Mathematics* 141: 553–572.

Thom, René (1975). *Structural Stability and Morphogenesis: An Outline of a General Theory of Models.* Reading: Benjamin.

Trigg, Charles W. (1978). What Is Recreational Mathematics? *Mathematics Magazine* 51: 18–21.

Turing, Alan (1936). On Computable Numbers with an Application to the Entscheidungs Problem. *Proceedings of the London Mathematical Society* 42: 230–265.

Tymoczko, Thomas (1979). The Four-Color Problem and Its Philosophical Significance. *Journal of Philosophy* 24: 57–83.

Uexküll, Jakob von (1909). *Umwelt und Innenwelt der Tierre.* Berlin: Springer.

Vajda, Steven (1989). *Fibonacci and Lucas Numbers, and the Golden Section.* Chichester: Ellis Horwood.

Verene, Donald P. (1981). *Vico's Science of Imagination.* Ithaca: Cornell University Press.

Vernadore, J. (1991). Pascal's Triangle and Fibonacci Numbers. *The Mathematics Teacher* 84: 314-316.

Visser, Beth A., Ashton, Michael C., and Vernon, Philip A. (2006.) g and the Measurement of Multiple Intelligences: A Response to Gardner. *Intelligence* 34: 507–510.

Vorderman, Carol (1996). *How Math Works.* Pleasantville: Reader's Digest Association.

Warner, George F. (ed.) (2015). *The Voyage of Robert Dudley Afterwards Styled Earl of Warwick & Leicester and Duke of Northumberland.* Amazon: Scholar's Choice.

Warnsdorf, H. C. von (1823). *Des Rösselsprunges einfachste und allgemeinste Lösung.* Schmalkalden: Varnhagen.

Watkins, John J. (2004). *Across the Board: The Mathematics of Chess Problems.* Princeton: Princeton University Press.

Wells, David (1992). *The Penguin Book of Curious and Interesting Puzzles.* Harmondsworth: Penguin.

Wells, David (2005). *Prime Numbers: The Most Mysterious Figures in Math.* Hoboken: John Wiley.

Wells, David (2012). *Games and Mathematics: Subtle Connections.* Cambridge: Cambridge University Press.

Wildgang, Wolfgang and Brandt, Per Aage (2010). *Semiosis and Catastrophes: René Thom's Semiotic Heritage.* New York: Peter Lang.

Wiles, Andrew (1995). Modular Elliptic Curves and Fermat's Last Theorem. *Annals of Mathematics. Second Series* 141: 443–551.

Willerding, Margaret (1967). *Mathematical Concepts: A Historical Approach.* Boston: Prindle, Weber & Schmidt.

Williams, Anne D. (2004). *The Jigsaw Puzzle: Piecing Together a History*. New York: Berkley Books.

Wilson, Robin (2002). *Four Colors Suffice: How the Map Problem Was Solved*. Princeton: Princeton University Press.

Wittgenstein, Ludwig (1922). *Tractatus Logico-Philosophicus*. London: Routledge and Kegan Paul.

Zadeh, Lofti A. (1965). Fuzzy Sets. *Information and Control* 8: 338–353.

Index

© Springer International Publishing AG, part of Springer Nature 2018
M. Danesi, *Ahmes' Legacy*, Mathematics in Mind,
https://doi.org/10.1007/978-3-319-93254-5

Printed in the United States
By Bookmasters